Sevak Khachadorian

Vibrational properties of silicon nanowires

Sevak Khachadorian

Vibrational properties of silicon nanowires

The effects of strain, temperature, and size on the vibrational properties of silicon nanowires

Südwestdeutscher Verlag für Hochschulschriften

Impressum/Imprint (nur für Deutschland/only for Germany)
Bibliografische Information der Deutschen Nationalbibliothek: Die Deutsche Nationalbibliothek verzeichnet diese Publikation in der Deutschen Nationalbibliografie; detaillierte bibliografische Daten sind im Internet über http://dnb.d-nb.de abrufbar.
Alle in diesem Buch genannten Marken und Produktnamen unterliegen warenzeichen-, marken- oder patentrechtlichem Schutz bzw. sind Warenzeichen oder eingetragene Warenzeichen der jeweiligen Inhaber. Die Wiedergabe von Marken, Produktnamen, Gebrauchsnamen, Handelsnamen, Warenbezeichnungen u.s.w. in diesem Werk berechtigt auch ohne besondere Kennzeichnung nicht zu der Annahme, dass solche Namen im Sinne der Warenzeichen- und Markenschutzgesetzgebung als frei zu betrachten wären und daher von jedermann benutzt werden dürften.

Verlag: Südwestdeutscher Verlag für Hochschulschriften GmbH & Co. KG
Heinrich-Böcking-Str. 6-8, 66121 Saarbrücken, Deutschland
Telefon +49 681 37 20 271-1, Telefax +49 681 37 20 271-0
Email: info@svh-verlag.de

Approved by: Berlin , TU , Diss. , 2007

Herstellung in Deutschland:
Schaltungsdienst Lange o.H.G., Berlin
Books on Demand GmbH, Norderstedt
Reha GmbH, Saarbrücken
Amazon Distribution GmbH, Leipzig
ISBN: 978-3-8381-2900-6

Imprint (only for USA, GB)
Bibliographic information published by the Deutsche Nationalbibliothek: The Deutsche Nationalbibliothek lists this publication in the Deutsche Nationalbibliografie; detailed bibliographic data are available in the Internet at http://dnb.d-nb.de.
Any brand names and product names mentioned in this book are subject to trademark, brand or patent protection and are trademarks or registered trademarks of their respective holders. The use of brand names, product names, common names, trade names, product descriptions etc. even without a particular marking in this works is in no way to be construed to mean that such names may be regarded as unrestricted in respect of trademark and brand protection legislation and could thus be used by anyone.

Publisher: Südwestdeutscher Verlag für Hochschulschriften GmbH & Co. KG
Heinrich-Böcking-Str. 6-8, 66121 Saarbrücken, Germany
Phone +49 681 37 20 271-1, Fax +49 681 37 20 271-0
Email: info@svh-verlag.de

Printed in the U.S.A.
Printed in the U.K. by (see last page)
ISBN: 978-3-8381-2900-6

Copyright © 2011 by the author and Südwestdeutscher Verlag für Hochschulschriften GmbH & Co. KG and licensors
All rights reserved. Saarbrücken 2011

I dedicate this work to my mother, Sonik Stephanian.

"Wish You Were Here."

Abstract

The building block of today's digital world is based on the silicon technology. Moreover, silicon is widely available, well studied and therefore will remain a very important part of digital technologies. In recent years the synthesis of semiconductor materials, structures and devices with dimensions ranging from a few angstrom to several nanometers, by a number of techniques [1], brings a new quality in the miniaturization of the semiconductor structures. Thin surfaces, nanowires, nanorods and quantum dots, show different mechanical, electrical and optical properties from those of the bulk Si [1–5]. An interesting aspect of confinement is that through varying of the size of the materials, their physical properties can be tuned. The deployment of one-dimensional nanostructures, such as silicon nanowires (SiNWs), in future generations of integrated electronic and optoelectronic devices makes the research on SiNWs promising and interesting [6].

The possible future application of SiNWs is conditional upon the studying of its mechanical, electrical and optical properties. Available works on mechanical properties of SiNWs show significant deviations from one another [7–12]. In this thesis the elastic properties, such as the bulk modulus and isothermal compressibility of SiNWs, were studied by means of high pressure Raman spectroscopy, accompanied by transmission electron microscopy (TEM). The findings showed a more pronounced pressure coefficient and consequently a higher compressibility for SiNWs compared to the bulk Si. Incorporating the resulted data obtained from Raman spectra of SiNWs under hydrostatic pressure and high-resolution TEM images, revealed that the lattice expansion of the Si-core was induced by its oxide shell. In a similar fashion, the effects of the oxide shell on the elastic and vibrational properties of SiNWs were studied. In core-shell structures, the coverage of a SiNW with a SiO_x shell leads to a tensile strain of the Si core. Calculations based on the linear continuum elasticity model confirmed this observation.

The remarkable increase of the FWHM of the Raman peak under external pressure, which was predicted theoretically to be a result of the LTO-phonon decay into LO+TA phonons, was observed in this study. This increase started at a lower pressure (4 GPa) for the SiNWs than the predicted value for the bulk Si (at about 7GPa). Moreover, the thickness of the oxide layer of SiNWs was found to affect the value of the critical pressure for the activation of a new decay channel.

The temperature-dependent Raman measurements of SiNWs showed that the three and four phonon anharmonic processes were more pronounced than their bulk counterparts. Moreover, the cross-section of the electron-phonon interactions in the case of the acoustic phonons was found to increase as the temperature ascended.

It was found that the observed saturation of Raman frequencies, measured on a bundle of SiNWs under increasing excitation power, was a consequence of morphological changes in SiNWs. These changes were determined to affect only the macroscopic and not the microscopic structure of SiNWs.

Contents

1	**Introduction**	**1**
2	**Si nanowires**	**5**
	2.1 Silicon	5
	2.2 Silicon nanowires growth	6
	2.3 Samples	9
	2.3.1 (8+3)-SiNWs	10
	2.3.2 (60+5)- and (60+20)-SiNWs	11
	2.3.3 (15+5)-SiNWs	15
3	**Raman scattering**	**17**
	3.1 Basics of Raman scattering	17
4	**Geometry dependence of the phonon modes in Si nanowires**	**23**
	4.1 The Phonon confinement: RCF-model	24
5	**Pressure dependence of the vibrational properties of SiNWs**	**29**
	5.1 Theory: Effect of pressure on the vibration properties of solids	30
	5.1.1 Pressure dependent force constant: Anharmonicity	30
	5.1.2 Elastic properties and Raman scattering	32
	5.1.3 Frequency shift of LTO-phonons under hydrostatic pressure	34
	5.2 Raman scattering of silicon nanowires under hydrostatic pressure	36
	5.2.1 Experimental setup	36
	5.2.1.1 Raman Spectrometer	36

		5.2.1.2	Diamond anvil cell (DAC)	37

- 5.2.2 Results and discussion 38
  - 5.2.2.1 Is there any heating effect? 38
  - 5.2.2.2 Raman shift under hydrostatic pressure 40
  - 5.2.2.3 Relation between Bulk modulus and lattice constant of SiNWs 44
  - 5.2.2.4 Raman linewidth of SiNWs under hydrostatic pressure 47
- 5.2.3 Conclusions 51
- 5.3 The effect of oxide layers on the vibrational properties of silicon nanowires 52
  - 5.3.0.1 Raman shift of (60+5)- and (60+20)-SiNWs under pressure 54
  - 5.3.0.2 Raman linewidth of (60+5)- and (60+20)-SiNWs under hydrostatic pressure 62
  - 5.3.0.3 Conclusions 63
- 5.4 Strain in Si-SiO$_x$ core-shell nanowires 64
  - 5.4.1 Strain in linear continuum elasticity model 65
  - 5.4.2 Transformation of stiffness tensor 67
  - 5.4.3 Strain and the strain energy in Si-SiO$_x$ core-shell nanowires . . 69
  - 5.4.4 Conclusion 75

6 Vibrational properties of SiNWs: temperature dependence   77

- 6.1 Theoretical approach: temperature-dependent Raman scattering . . . 78
  - 6.1.1 Balkanski model 79
  - 6.1.2 Cui model 80
  - 6.1.3 Mishra model 80
- 6.2 Raman scattering measurements of SiNWs: temperature dependence . 81
  - 6.2.1 Experimental details 81
  - 6.2.2 Results and discussion 83
    - 6.2.2.1 Interpretation of the results based on Balkanski *et al.* and Cui *et al.* models 86
    - 6.2.2.2 Temperature dependence of the 2TA(X) Raman band 90
    - 6.2.2.3 Temperature dependence of the 2TO Raman band . 91

		6.2.2.4	The temperature effect on the intensities of 2TA(X) and 2TO bands	95

6.3 Conclusion 99

7 Excitation power dependence on the vibrational properties of SiNWs 101
7.1 Experimental details 102
7.2 Results 102
    7.2.1 Excitation power dependence measurements in ambient conditions 103
    7.2.2 Excitation power dependence measurements (in vacuum) . . . 105
    7.2.3 Homogeneous temperature dependent measurements 108
7.3 Conclusion 108

8 Summary and outlook   111

A Appendix   115
A.1 Murnaghan equation of state 115
A.2 Brich Murnaghan equation of state 116

Bibliography   117

The most exciting phrase to hear in science, the one that heralds new discoveries, is not "Eureka!" (I found it!) but "That's funny".

Isaac Asimov

1

Introduction

Today's information technology is based on silicon. Materials with dimensions ranging from a few angstroms to several nanometers can be routinely synthesized by a number of techniques [1]. Owing to their confined size and high surface-to-volume ratio, these can show different mechanical, electrical and optical properties from those of the bulk Si [1, 4, 5, 13, 14]. Aside from potential technological applications, they provide ample opportunity in the fundamental research field, for instance to investigate size- and shape-dependent confinement effects. Spheres, cubes [15], rods [16], disks [17], wires [18], tetrapods [19] and arrows [20] can be prepared from a wide variety of materials. Nanostructured silicon is particularly interesting, because today's semiconductor technology is still largely pinned on this material. Silicon nanowires (SiNWs) have stimulated extensive efforts, ranging from the integration of optoelectronic devices into Si microelectronics [21–23], to large-area applications such as photovoltaics [24–26] and thermoelectrics [27, 28].

Chapter 1. Introduction

Raman spectroscopy has proven to be an effective and nondestructive characterization technique to understand the lattice dynamics of SiNWs [29–35]. Phonon frequencies, linewidths and intensities give valuable information about microscopic parameters such as bonding and structure as well as deviations from their crystalline counterparts [14, 36, 37]. Incorporating the data obtained by Raman spectroscopy with those gained *via* transmission electron microscopy (TEM), such as the diameter and the lattice parameter (expansion or contraction) results in a better understanding of the elastic, thermal and vibrational properties of SiNWs. The SiNWs studied in this work were analyzed by means of high resolution TEM, with results presented in Chapter 2. These measurements provided valuable information regarding the micro- and macro-structural properties of SiNWs.

In Chapter 3.1 an introduction to the theory of the Raman scattering, including the first-order and second-order scattering, are presented. The optical phonons in SiNWs are confined to the SiNWs volume. The confinement of a phonon wavefunction leads to a relaxation of the $\vec{q} = 0$ rule, and consequently to the frequency shift and broadening of phonon band of SiNWs. This issue is discussed in Chapter 4 using the confinement model of Ricther textitet al. [38], and the Campbell and Fauchet model (RCF) [39].

Employment of SiNWs in the future technologies demands a very detailed investigation of these nanostructures under applied stress and in different temperatures. In Chapter 5.2 the effect of external hydrostatic stress on the vibrational properties of SiNWs with a core diameter of 8 nm are presented. The pressure coefficient of the first-order optical Raman mode of SiNWs were determined. The results obtained from the effect of applied hydrostatic pressure on Raman frequencies as well as on FWHM of SiNWs provided information about the elastic properties of SiNWs and the decay process of optical phonons.

SiNWs are typically covered with a SiO_x shell with variable thicknesses. The effect of oxide shell on elastic properties of SiNWs was studied in Chapter 5.3. Two SiNW samples with the same core diameter of 60 nm, but different SiO_x shell thicknesses were employed to study the effect of oxide shell on vibrational and elastic properties of SiNWs. In order to confirm and explain the experimental results, the elastic continuum model was used. The effect of temperature on the properties of SiNWs was also a very important component of this study concerning a possible future application, which is presented in Chapters 6 and 7. In particular, Chapter 6 deals with the effect of homogeneous heating on the vibrational properties of SiNWs, including the temperature effect on the first-order as well as second-order Raman scattering of SiNWs. The second-order Raman scattering measurements are essential, because in higher order Raman scattering the optical phonons at the Γ point, as well as the phonons with larger wave vector participate in the scattering process, giving additional information related to the vibrational properties of SiNWs. Moreover, the study of the intensities of the Raman modes and features as a function of temperature in SiNWs provides information about the electron-phonon interaction.

In Chapter 7, the effect of inhomogeneous temperature on the first-order Raman mode of SiNWs is presented, exhibiting different results from previous experiments, which were conducted at homogeneous temperatures (Chapter 6).

Science cannot solve the ultimate mystery of nature. And that is because, in the last analysis, we ourselves are a part of the mystery that we are trying to solve.

Max Planck

2

Si nanowires

2.1 Silicon

Silicon (Si) is one of the most important semiconductor materials which impacts our daily life. The phase of Si which is stable in ambient conditions exhibits diamond cubic crystal structure (see Figure 2.1). This consists of two tetrahedrally bonded atoms in each primitive cell, which are separated by 1/4 of the width of the unit cell in each dimension. The atoms are ordered in the diamond structure by covalent sp^3 bonds. Increasing the external pressure, causes Si to transfer to β-Si at 12 GPa, to the intermediate orthorhombic phase at about 13 − 16 GPa, to the primitive hexagonal (ph) structure at about 16 GPa, and finally to the hexagonal close-packed (hcp) structure at above 42 GPa [40]. Diamond cubic structure of silicon has a lattice constant of $a = 0.5430710$ nm. Si has indirect electronic bandgap of 1.12 eV as shown in Figure 2.2. The minimum of the conduction band is at 85% of the distance

Chapter 2. Si nanowires

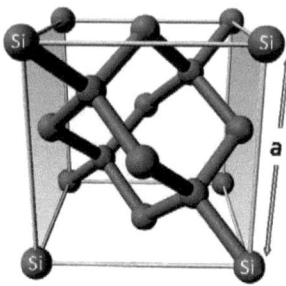

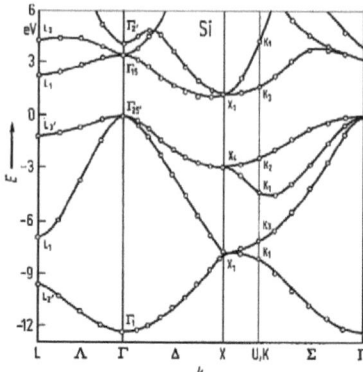

Figure 2.1: Diamond-like crystal structure of Si. The primitive basis of Si structure has two atoms at fractional coordinates of (000) and (1/4 1/4 1/4). The face-centered cubic structure has a lattice spacing of $a = 0.5430710$ nm.

Figure 2.2: Silicon band structure, taken from Raference [41]. Solid lines are the data from non-local energy dependent pseudooptical calculation and the circles are calculated by localized atomic orbital method.

between the Γ-point and the X-point in the Δ-direction. In Raman scattering, which will be discussed in Chapter 3.1, the interaction of the laser light and the phonons is mediated by the electronic system. The laser light absorption takes place through direct bandgap. The minimal direct band for Si is about 3.4 eV (between the valence band Γ'_{25} and the conduction band Γ_{15}).

2.2 Silicon nanowires growth

In order to synthesize silicon nanowires (SiNWs), different techniques have been developed in the past. The different growth techniques differ mainly in the way Si is supplied and with respect to the catalyst material. In some of these methods the wires are grown directly by elemental Si, whereas in others Si is provided as a silicon compound. Some of these techniques are:

1. **Chemical Vapor Deposition (CVD)**: In CVD technique Si is provided by an oxygen-free precursor, such as silane (SiH_4), disilane (Si_2H_6) and dichlorosilane (H_2SiCl_2). The growth of SiNWs in this technique can be carried out at high (800° C) or low (< 700° C) temperatures.

    - **High Temperature CVD:** The growth of SiNWs using high temperature CVD is performed in tubular hot-wall reactors [42, 43]. The carrier gas, which is hydrogen or a hydrogen/inert gas mixture, flows through an externally heated quartz tube. The pressure is held at about atmospheric pressure. Prior to entering the reactor, part of the gas is led through a bubbler filled with $SiCl_4$, which is liquid at ambient condition. The Si sample, already deposited with some amount of the catalyst metal, is placed in the hot zone of the reactor. As $SiCl_4$ is supplied to the reactor, the SiNWs begin to grow. The main advantage of high temperature CVD is the much broader range of potential Vapor-Liquid-Solid (VLS) catalyst materials. Gold (Au) and copper (Cu) [42] yield excellent results at temperatures above 850° C. At even higher temperatures, platinum (Pt) and nickel (Ni) are also good choices.

    - **Low Temperature CVD:** Low temperature CVD operates at temperatures below 700° C. The typical precursor for low temperature CVD is SiH_4. In low temperature CVD technique, a plasma can be used to pre-crack the Si precursor in order to enhance the growth effectiveness of the SiNWs [44, 45]. The use of a strong electric field is also known to enhance the growth process of SiNWs [46, 47].

2. **Thermal evaporation of silicon monoxide (SiO):** Another prominent synthesis approach is based on the thermal evaporation of SiO at about 1100 − 1350° C. An inert gas supply is connected to a tube furnace. For the successful synthesis of SiNWs, it is important to have a temperature gradient in the tube furnace so that the inert gas flows from the hotter to the colder part of the furnace. SiO is placed in the hotter zone of the tube furnace, where it evaporates. The evaporated SiO is carried away by the gas stream to the cooler end of the tube, resulting in a disproportionation reaction into Si and

Chapter 2. Si nanowires

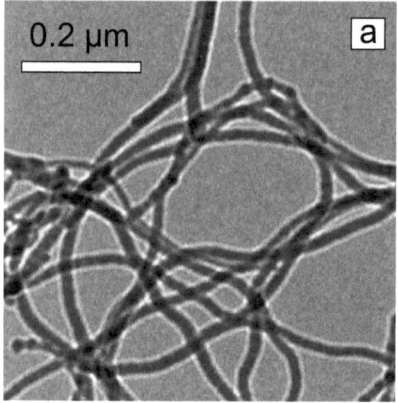 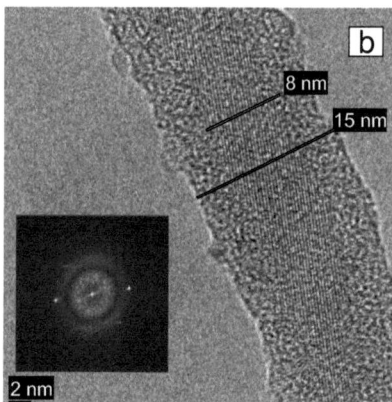

Figure 2.3: Representative morphology of the (8+3)-SiNWs sample.

Figure 2.4: Individual SiNW and its diffraction pattern (inset).

SiO_2 (at about $1100 - 1350°$ C) and subsequent growth of SiNWs [48]. These SiNWs are normally covered by a thick SiO_x shell ($1.5 \leq x \leq 2.8$).

3. **Laser Ablation:** The growth of SiNWs by laser ablation is a simple technique, which is performed in a tube furnace. The laser ablation target (*e.g.* a mixture of Si-Fe, containing about 90% Si and 10% Fe), is placed into the tube furnace. The tube furnace is heated to a temperature of 1200° C. Argon gas flows through the furnace and the pressure is held at 500 Torr. Fe and Si are ablated from the target using a pulsed high power laser. The ablated material hits the inert gas molecules and condenses, forming Fe-Si nanodroplets, which are components of the SiNWs growth-process via VLS. The generated SiNWs can be scraped from the reactor.

4. **Molecular Beam Epitaxy (MBE):** The SiNWs growth via MBE is achieved by evaporating Si onto a catalyst covered substrate, typically Si(111). In this method a ultra-high vacuum (UHV) system (10^{-10} mbar) is used to prevent oxidation or contamination of the substrate or the SiNWs. Before Si evaporation, Au is deposited onto the substrate. Annealing the substrate at temperatures

2.3 Samples

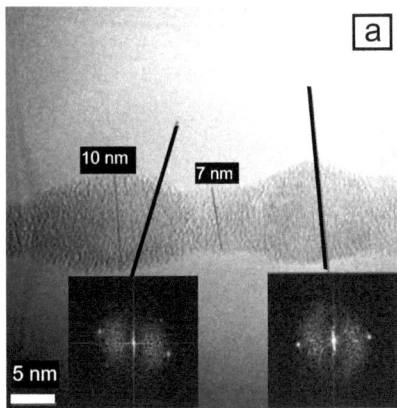

Figure 2.5: TEM image of Si-SiO$_2$ nanochains with different core orientation refer to the wire growth direction.

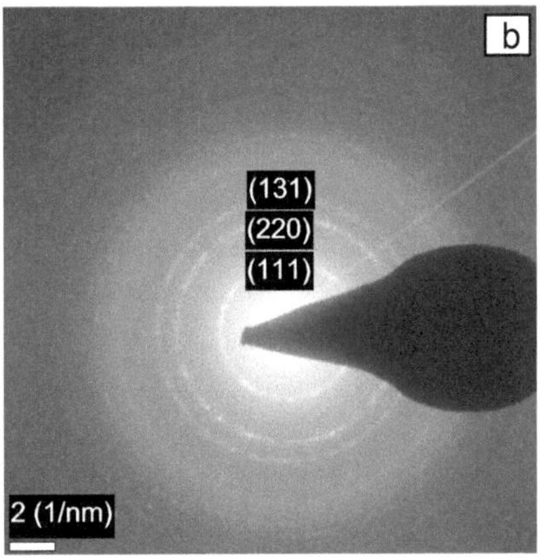

Figure 2.6: Diffraction pattern of the SiNWs presented in Figure 2.3. This proves that the SiNWs grow along different crystallographic directions, and the growth directions are (111), (220) and (131).

above the Au-Si eutectic temperature results in formation of an Au-Si alloy, which act as catalysts for SiNWs growth.

The SiNWs studied in this thesis are synthesized by catalytic chemical vapor deposition (CVD) and high temperature thermal evaporation techniques.

2.3 Samples

In this section, the samples studied in the present work are introduced. The samples are labeled as (X+Y)-SiNWs, where X and Y are the diameter of silicon core and the thickness of the oxide layer, respectively.

Chapter 2. Si nanowires

2.3.1 (8+3)-SiNWs

SiNWs were grown by high-yield vapor transport [49] as described before. In this method, substrate-bound SiNWs were grown by Au-assisted deposition in a vapor-transport reactor. Briefly, SiO powder was evaporated at about 1400° C in a horizontal tube furnace for 3 hours. The vapor condensed at about 900° C on a quartz substrate to form SiNWs. During the synthesis process, Ar was allowed to flow (100 sccm) as carrier gas at pressures close to atmosphere (800 − 1000 mbar). As-grown SiNWs were then sonicated in isopropanol and dispersed on high-grade steel. The initial characterization of the sample was performed with 200 kV transmission electron microscope (TEM) (Tecnai G^2 20 from FEI). The average wire diameter was 15 nm, consisting of an outer SiO_2 shell of 3 − 5 nm and a crystalline silicon core. The SiNWs were tens of microns in length. Figure 2.3 shows a representative TEM image of the SiNWs. These have a core diameter of ~ 8 nm and an amorphous SiO_2-coating of ~ 3.5 nm, as shown in Figure 2.4. The inset of Figure 2.4 plots the Fourier-transform of the crystallographic planes of an individual SiNW. This Fourier-transformed image indicates that the crystallographical growth direction is unchanged along the presented SiNW, as a change would be characterized by more reflections in the image. From the diffraction pattern (inset of Figure 2.4), the lattice parameter (a) is determined to be 5.57(5) Å, which shows an expansion of $\sim 2\%$, compared to the bulk Si (5.43 Å [50]). This is consistent with previous X-ray diffraction (XRD) studies of SiNWs, where a higher lattice parameter value was reported compared to the bulk Si (0.1% [51], 0.4% [14], or 0.31% [52]). However, others found a lower value using TEM imaging (0.14%) [53], (see Table 5.2).

Reducing the pressure to about 400 mbar during the SiNWs growth has proven to enhance the formation of the Si-SiO_2 nanochains (*i.e.* filamentary nanostructures where crystalline Si spheres are connected by SiO_2 bridges of variable length) [49, 54]. In case of (8+3)-SiNWs, in order to minimize the formation of Si-SiO_2 nanochains, the carrier gas (Ar) was flowed at pressures about 800 − 1000 mbar. Nevertheless as shown in Figure 2.5, some Si-SiO_2 nanochains were observed in TEM images. They consist of crystalline cores with a diameter of about 7 nm separated by amorphous SiO_2 regions. The SiO_2-shell is about 1.5 nm thick. The crystallographic planes in

2.3 Samples

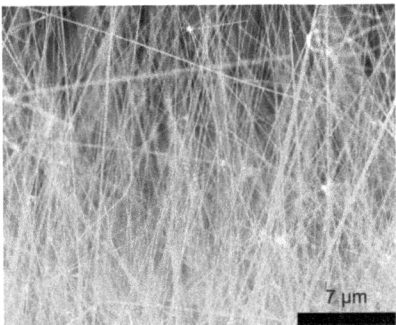

Figure 2.7: SEM image of (60+5)-SiNWs sample. The mean diameter of these SiNWs is about 60 nm and they are up to a few microns long.

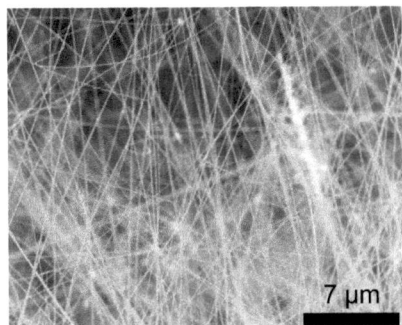

Figure 2.8: SEM image of (60+20)-SiNWs sample. The mean diameter of these SiNWs is about 80 nm (oxide shell included) and they are up to a few microns long.

each core are highlighted by black solid lines. The Fourier-transformation of the core images, shown in the inset of Figure 2.5, indicates that the crystalline regions have different growth orientation. By analyzing various TEM images of (8+3)-SiNWs, one can infer that the ratio of the Si-SiO$_2$ nanochains to the SiNWs is about 10%. Figure 2.6 shows the diffraction pattern of the SiNWs sample, from which the growth directions of (8+3)-SiNWs are identified as (111), (220) and (131).

2.3.2 (60+5)- and (60+20)-SiNWs

The (60+5)- and (60+20)-SiNWs were synthesized by catalytic chemical vapor deposition (CVD), using silane as precursor and Au as catalyst. Thin Au films (99.99%) were evaporated onto n-type Si(100) with a 200 nm-thick SiO$_2$ layer. The thickness of the Au film was 2 nm for (60+5)-SiNWs and 0.5 nm for (60+20)-SiNWs. The film thickness was monitored in-situ by a quartz crystal thickness and calibrated ex-situ by spectroscopic ellipsometry. In the stainless steel CVD chamber (base pressure below 7 − 10 mbar) the samples were placed on a pyrolytic boron nitride heater stage and typically pre-annealed in 1 mbar H$_2$ for 15 minutes at 380° C. Silane was then

Chapter 2. Si nanowires

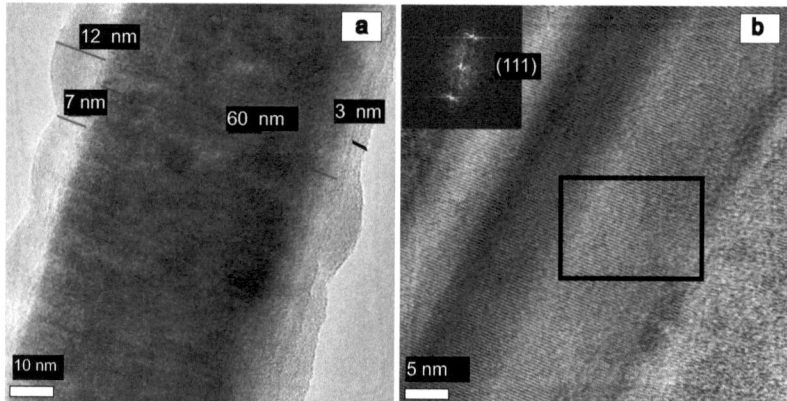

Figure 2.9: Structural characterization of (60+5)-SiNWs. (a) TEM image of an individual SiNW with a diameter of about 60 nm and an amorphous SiO_2 layer ranging between 3 and 12 nm. (b) TEM image of a single SiNW, (111) lattice planes and the diffraction pattern (inset) of the specified region (black box).

introduced to reach a growth pressure of 7 mbar. The growth time was 10 minutes. Figure 2.7 shows a scanning electron microscopy (SEM) image of (60+5)-SiNWs. As shown, these wires are several microns in length and are in contact with each other, but well separated from the substrate.

Figure 2.9(a) shows a high resolution TEM image of a representative single SiNW from the (60+5)-SiNWs sample. These nanowires had an average diameter of about 60 nm and were, in general, found to be crystalline with $3-12$ nm thick SiO_x coating, as shown in Figure 2.9(a). The diffraction images revealed that the growth of these SiNWs is along the (111) direction (Figure 2.9(b)). From the diffraction image of a selected region of the lattice planes shown in inset of Figure 2.9(b), the lattice constant of (60+5)-SiNWs was determined to be $a = 5.44(5)$ Å. This value exhibits 0.3% deviation from that of its bulk counterpart ($a = 5.431$ Å).

The (60+20)-SiNWs are grown in the same conditions as the (60+5)-SiNWs sample. The only difference is that the thickness of the Au film used in growth of (60+20)-

2.3 Samples

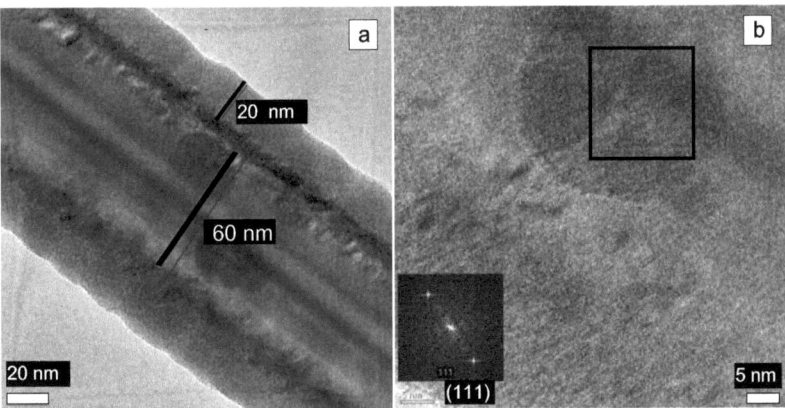

Figure 2.10: Structural characterization of (60+20)-SiNWs. (a) TEM image of an individual (60+20)-SiNW with a diameter of about 60 nm and a shell consisting SiO_x layer and amorphous Si of about 20 nm. (b) TEM image of single SiNW, (111) lattice planes and the diffraction pattern (inset) of the specified region (black box).

SiNWs, is 0.5 nm. As shown in Figure 2.8, the nanowires in this sample are also a few micrometer long and their diameters are almost the same as those of the (60+5)-SiNWs. Comparing the SEM images of these two samples reveal that the nanowire density of the (60+5)-SiNWs is higher than that of (60+20)-SiNWs. As it will be exhibited in Chapter 5.3 the lower density of (60+20)-SiNWs is reflected in the weaker Raman scattering signal of these nanowires compared to (60+5)-SiNWs. Figure 2.10 shows a high resolution TEM image of a single SiNW from the (60+5)-SiNWs sample. The average diameter of these SiNWs is about 60 nm. The shell covering these nanowires, which consists of oxide layer (SiO_x) and amorphous Si was found to be about 20 nm, as shown in Figure 2.10(a). Due to the thickness of SiO_x and the existing amorphous Si layer, it was very difficult to observe the lattice planes of the crystalline part. Figure 2.10(b) exhibits the lattice planes, which are oriented in (111) direction. This was revealed by the diffraction image of a selected area of the lattice planes (see the inset of Figure 2.10(b)).

Chapter 2. Si nanowires

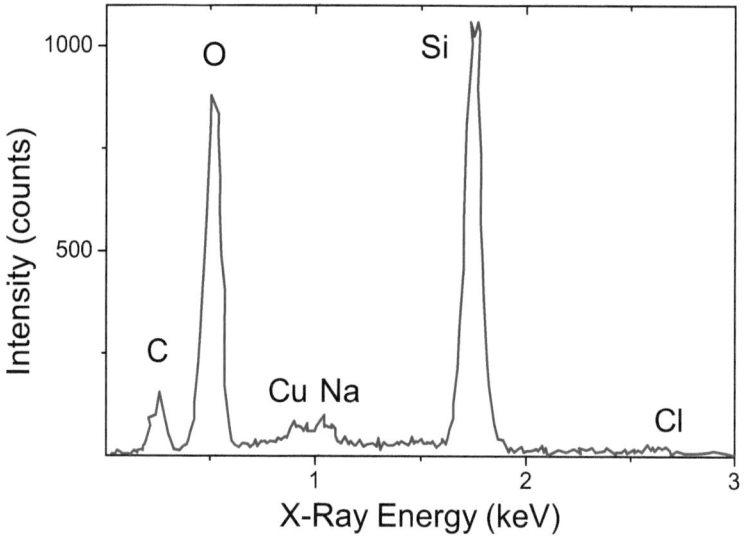

Figure 2.11: The energy dispersive X-ray (EDX) spectrum of (15+5)-SiNWs for 5 keV electrons.

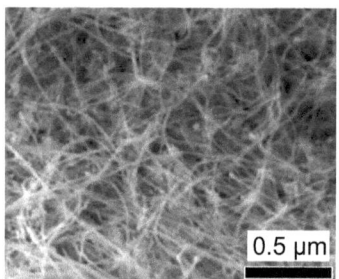

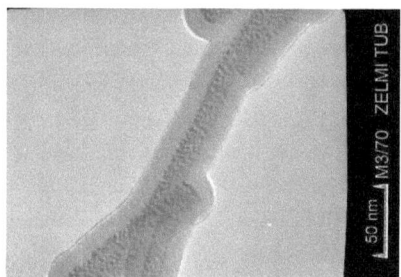

Figure 2.12: SEM image of (15+5)-SiNWs sample. The mean diameter of these SiNWs is about 15 nm and they are up to a few microns long.

Figure 2.13: TEM image of (15+5)-SiNWs.

2.3.3 (15+5)-SiNWs

The (15+5)-SiNWs were grown at temperatures below 400° C by plasma enhanced chemical vapor deposition, using silane as the Si source and Au as the catalyst [34]. After the deposition a layer of SiNWs was left on the reactor walls. The average wire diameter was 15 nm, coated with an outer SiO_2 shell of 5 nm [55]. The SiNWs were then dissolved in isopropanol. Depending on the volume of isopropanol, various concentrations of SiNWs could be obtained. The solvent was dribbled on a substrate, establishing a small number (2−5) of SiNW spots. Each sample spot was 1−2 mm in diameter and approximately $20-50\,\mu$m thick. The substrate was a polished copper bar with an additional silver coating. Figure 2.11 shows the energy dispersive X-ray (EDX) spectra of (15+5)-SiNWs measured with a Hitachi S-2700 SEM equipped with a SAMx EDX-System. The characteristic peaks of various elements are apparent in the Figure 2.11. Aside from two large peaks, which are associated with Silicon and oxygen, some other elements like carbon, copper, sodium and chlorine are also present. Figure 2.12 shows the SEM image of SiNWs taken by a Hitachi S-2700 SEM. The SiNWs are randomly arranged and their length is several micrometers. Moreover, the SiNWs are free-standing and have only few contact points with the other SiNWs. Figure 2.13 shows a TEM image of a single SiNW. The diameter of this SiNW is above the average diameter of 15 nm, determined by the group of Andrea Ferrari and John Robertson (Department of Engineering, University of Cambridge). In this figure, the border between the dark and light parts of the nanowire is not necessarily the accurate interface of $Si-SiO_x$. This is due to a typical diffraction feature, which results from an imperfectly focused electron beam.

Measure what can be measured, and make measureable what cannot be measured.

Galileo Galilei

3

Raman scattering

3.1 Basics of Raman scattering

Raman spectroscopy is one of the most important methods for investigating vibrational modes of crystalline materials, and has proven to be a powerful, quick and non destructive characterization technique. Raman spectroscopy measures the inelastic scattering of light in a sample via its elementary excitation. The incident light, with the electrical field of $\vec{E}(\vec{r},t)$, induces polarization $\vec{P}(\vec{r},t)$ in the sample, proportional to its the electric susceptibility χ:

$$\vec{P}(\vec{k}_i,\omega_i) = \chi(\vec{k}_i,\omega_i)\vec{E}_i(\vec{k}_i,\omega_i) \tag{3.1}$$

The electric susceptibility χ is in general a second rank tensor, though can be represented by a scalar in isotropic medium. A medium at a finite temperature has

Chapter 3. Raman scattering

a fluctuated electric susceptibility (χ), due to thermally exited atomic vibrations. These fluctuations are quantized into phonons and can be expressed as plane wave displacements:

$$\vec{Q}(\vec{r},t) = \vec{Q}(\vec{q},\omega_0)\cos(\vec{q}\cdot\vec{r}-\omega_0 t) \tag{3.2}$$

where q and ω_0 are the wavevector and frequency, respectively. These vibrations modify the electrical susceptibility χ. As $\vec{Q}$ displacements are small compared to the lattice parameter, χ can be expanded as a Taylor series in $\vec{Q}(\vec{r},t)$:

$$\chi(\vec{k}_i,\omega_i,\vec{Q}) = \chi_0(\vec{k}_i,\omega_i) + \sum_m \left(\frac{\partial \chi}{\partial Q_m}\right)_0 \vec{Q}_m(\vec{r},t) + \tag{3.3}$$
$$\frac{1}{2}\sum_{m,n}\left(\frac{\partial^2 \chi}{\partial \vec{Q}_m \cdot \partial \vec{Q}_n}\right)\left(\vec{Q}_m \cdot \vec{Q}_n\right) + \cdots,$$

where χ_0 is the electrical susceptibility of the medium with no fluctuation. Incorporating Equation 3.4 into Equation 3.1 yields:

$$\vec{P}(\vec{r},t,\vec{Q}_m) = \vec{P}_0 + \vec{P}_1 + \vec{P}_2 \tag{3.4}$$

where

$$\vec{P}_0(\vec{r},t) = \chi_0(\vec{k},\omega_i)\vec{E}_i(\vec{k},\omega_i)\cos(\vec{k}_i\cdot\vec{r}-\omega_i t) \tag{3.5}$$

and

$$\vec{P}_1(\vec{r},t,\vec{Q}) = \frac{1}{2}\left(\frac{\partial \chi}{\partial Q}\right)_0 \vec{Q}(\vec{q},\omega_0)\vec{E}_i(\vec{k}_i,\omega_i) \times \tag{3.6}$$
$$\left\{\cos[(\vec{k}_i+\vec{q})\cdot\vec{r}-(\omega_i+\omega_0)t] + \cos[(\vec{k}_i-\vec{q})\cdot\vec{r}-(\omega_i-\omega_0)t]\right\}$$

The scattered light consists of the Rayleigh scattering ($\vec{P}_0$), Anti-Stokes inelastic scattering (first part of $\vec{P}_1$) with wavevector $\vec{k}_{AS} = (\vec{k}_i+\vec{q})$ and frequency of $\omega_{AS} = (\omega_i+\omega_0)$ and a Stokes inelastic scattering with wavevector $\vec{k}_S = (\vec{k}_i-\vec{q})$ and frequency of $\omega_S = (\omega_i-\omega_0)$ (second part of $\vec{P}_1$). The term $\vec{P}_2$ corresponds to the second-order Raman scattering that will be discussed *vide infra*.

$$\hbar\omega_s = \hbar\omega_i + \hbar\omega \tag{3.7}$$

3.1 Basics of Raman scattering

The wavevector and frequency are conserved in this scattering process. As a result, in case of the one phonon Raman scattering, the wavevector can not exceed $2k_i$. Consequently, only the phonons from the center of the Brillouin zone can participate in the one-phonon Raman scattering, if the visible laser light is used for excitation. The susceptibility χ_{ij} can be written as:

$$\chi_{ij} = (\chi_{ij})_0 + \sum_k \left(\frac{\partial \chi_{ij}}{\partial \vec{Q}_m}\right)_0 \vec{Q}_m + \cdots, \qquad (3.8)$$

where i and j indices are the space coordinates, and $\vec{Q}_m$ is the vector displacement of a given atom induced by the phonon. The intensity of the scattered light depends on the polarization of the scattered radiation $\hat{e}_s$ as $\left|\vec{P}_1 \cdot \hat{e}_s\right|^2$. Using Equation 3.6, the intensity of the scattered light can be written as $I_s \propto |\hat{e}_i \cdot R \cdot \hat{e}_s|^2$, where $\hat{e}_i$ is the polarization of the incident light and R is the Raman tensor, which is defined as $R = \left(\frac{\partial \chi}{\partial \vec{Q}}\right)_0 \hat{Q}_m(\omega_0)$, where $\hat{Q}_m = \vec{Q}_m/|\vec{Q}_m|$, and is obtained by the contraction of $\vec{Q}_m$ and taking the derivative of χ with respect to $\vec{Q}_m$. The Raman tensor regulates the Raman selection rules *i.e.* it determines which combination of e_i, e_s and the symmetry of the crystal structure, which affects the symmetry of the susceptibility tensor, results in a non-zero intensity of the scattered light.

For Si, the Raman tensors can be written as:

$$R(x) = \begin{pmatrix} 0 & 0 & 0 \\ 0 & 0 & d \\ 0 & d & 0 \end{pmatrix} \quad R(y) = \begin{pmatrix} 0 & 0 & d \\ 0 & 0 & 0 \\ d & 0 & 0 \end{pmatrix} \quad R(z) = \begin{pmatrix} 0 & d & 0 \\ 0 & 0 & 0 \\ d & 0 & 0 \end{pmatrix} \qquad (3.9)$$

The second-order Raman scattering can be calculated from the third term of Equation 3.4. In second-order Raman scattering two phonons participate (Equation 3.10), where generation and absorption of both, or generation of one and simultaneous absorption of the other occur. Unlike the first-order Raman scattering, where only phonons from Brillouin zone center participate, in second-order Raman scattering the phonons from the entire Brillouin zone can be involved. The normal coordinates of the vibration associated with frequencies of ω_{0m} and ω_{0n} are $\vec{Q}_m$ and $\vec{Q}_n$, respectively and can be written as:

Chapter 3. Raman scattering

$$\vec{Q}_m(\vec{r},t) = \vec{Q}_m(\vec{q}_m,\omega_{0m})\cos(\vec{q}_m\cdot\vec{r} - \omega_{0m}t)$$

$$\vec{Q}_n(\vec{r},t) = \vec{Q}_n(\vec{q}_n,\omega_{0n})\cos(\vec{q}_n\cdot\vec{r} - \omega_{0n}t)$$

(3.10)

Incorporating Equation 3.10 into the combination of Equations 3.1 and 3.4 yields the polarization in the case of the second-order Raman scattering:

$$\begin{aligned}\vec{P}_2(\vec{r},t,\vec{Q}_m,\vec{Q}_n) = & \frac{1}{8}\left(\frac{\partial^2\chi}{\partial\vec{Q}_m\cdot\partial\vec{Q}_n}\right)\left(\vec{Q}_m\cdot\vec{Q}_n\right)\vec{E}_i(\vec{k}_i,\omega_i) \cdot \\ & \Big[\cos[(\vec{q}_m+\vec{q}_n-\vec{k}_i)\cdot\vec{r} - (\omega_{0m}+\omega_{0n}-\omega_i)t] + \\ & \cos[(\vec{q}_m+\vec{q}_n+\vec{k}_i)\cdot\vec{r} - (\omega_{0m}+\omega_{0n}+\omega_i)t] + \\ & \cos[(\vec{q}_m-\vec{q}_n-\vec{k}_i)\cdot\vec{r} - (\omega_{0m}-\omega_{0n}-\omega_i)t] + \\ & \cos[(\vec{q}_m-\vec{q}_n+\vec{k}_i)\cdot\vec{r} - (\omega_{0m}-\omega_{0n}+\omega_i)t]\Big]\end{aligned}$$

(3.11)

As shown in Equation 3.11, there are four possibilities for the second-order Raman scattering, which cause a frequency shift of $\pm\omega_{0m}\pm\omega_{0n}$ in induced polarization. The Raman features resulted from two different phonons with frequencies of $\omega_{0m}+\omega_{0n}$ and $\omega_{0m}-\omega_{0n}$ are called combination and difference modes, respectively. In case of identical phonon-energies the resulted Raman peak is called the overtone.

Figure 3.1 shows the dispersion relation of bulk Si, taken from Reference [41]. The optical phonon modes in Si are degenerated at Γ-point. The energy of these branches corresponds to optical first-order Raman peak ($520\,\mathrm{cm}^{-1}$), which is noted as LTO(Γ) in Raman spectrum of Si shown in Figure 3.2. The symmetry of Si crystal governs the shape of dispersion curve in the certain high-symmetry regions of the Brillouin zone (BZ). The components of the energy gradient in k-space, at Γ, L, X and W points (critical points), must be equal to zero in order to have a high contribution to the phonon density-of-state. The phonon density-of-state exhibits some intense peaks at phonon energies, which are near to the critical point energies. Moreover, since the two-phonon Raman spectrum reflects the phonon density-of-state, in the

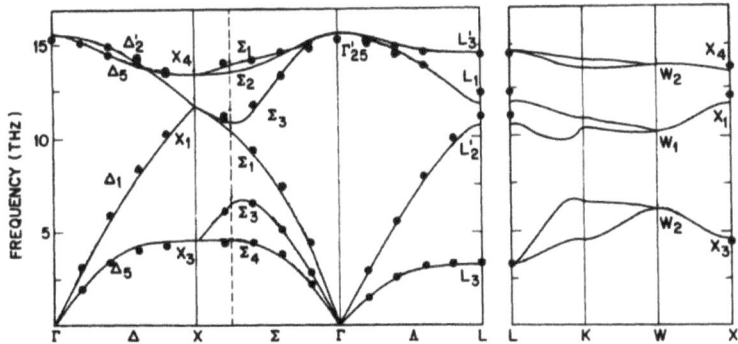

Figure 3.1: *Ab-initio* calculated phonon dispersion relation for the bulk Si, taken from Reference [41].

Raman Spectrum the contribution of various critical points to the observed Raman features can be identified.

Some of these features are shown in Figure 3.2. The peak observed at about 300 cm$_{-1}$ is related to the second-order transversal acoustic phonon contribution at the X-point. The feature observed in the 900-1100 cm$_{-1}$ corresponds to the contributions related to the second-order transversal optical phonons at Q, $S1$, W, L and Γ-points. In Chapter 6, these features and their temperature-dependence will be discussed in detail.

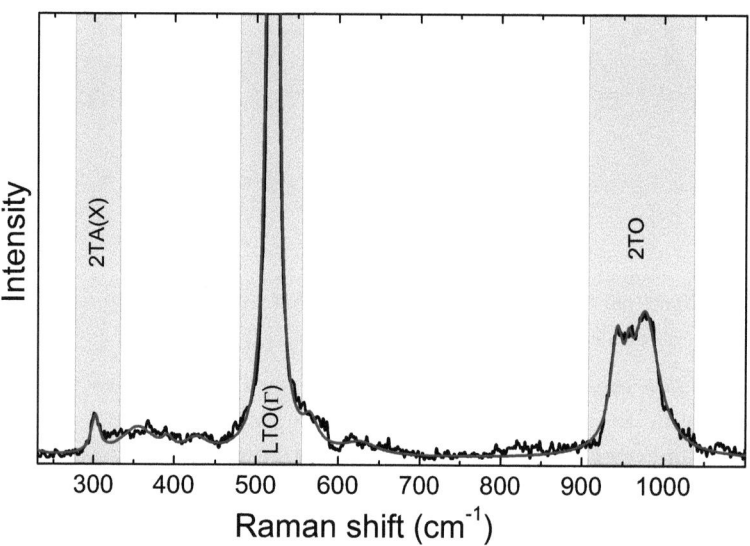

Figure 3.2: Raman spectrum of bulk Si measured with $\lambda = 514\,\text{nm}$ line of Ar^+ laser.

If all of mathematics disappeared, physics would be set back by exactly one week.

Richard P. Feynman

4
Geometry dependence of the phonon modes in Si nanowires

It is known that the first optical Raman mode of SiNWs is shifted compared to that of the bulk silicon. On the one hand, the compressive (tensile) stress in the crystal results in a Raman frequency shift towards the higher (lower) frequencies, which is known as blueshift (redshift). On the other hand, in the size restricted crystals such as SiNWs a redshift and a broadening of Raman peaks are observed. To avoid the misinterpretation of the observed Raman spectra of SiNWs, one must be able to distinguish the stress-related Raman shifts from those related to the size-effect. This chapter describes the confinement model of Ricther *et al.* [38], and the Campbell and Fauchet model (RCF) [39].

Chapter 4. Geometry dependence of the phonon modes in Si nanowires

4.1 The Phonon confinement: RCF-model

According to the Heisenberg uncertainty principle, the fundamental $q \approx 0$ Raman selection rule is relaxed for a finite size crystal. This allows the phonons away from the Brillouin zone center to take part in the scattering process. Assuming that the diameter of a nanostructure is d, the phonon uncertainty will be equal to $\Delta q \approx 1/d$. A general model for phonon confinement was first introduced by Richter et al. and applied for low-dimensional silicon structures [38]. This model was further specialized for nanowires by Campbell and Fauchet [39].

According to Richter et al. the wavefunction of a phonon with the wavevector $\vec{q_0}$ is restricted to the volume of the infinite crystal and can be written as:

$$\Phi(\vec{q_0}, \vec{r}) = u(\vec{q_0}, \vec{r}) \exp(-i\vec{q_0} \cdot \vec{r}) \qquad (4.1)$$

$u(\vec{q_0}, \vec{r})$ has the periodicity of the crystal lattice. In order to simplify the calculation, Richter et al. assumed that the crystal is spherical and the localized wavefunction has the form of a Gauss-distribution. Replacing the wavefunction Φ with $|\Psi| = A^2 \exp\left(-r^2/(L/2)^2\right)$ yields:

$$\begin{aligned} \Psi(\vec{q_0}, \vec{r}) &= W(\vec{q_0}, \vec{r}) \Phi(\vec{q_0}, \vec{r}) \\ &= A \cdot \exp\left(-(r^2/2)/(L/2)^2\right) \Phi(\vec{q_0}) , \\ &= \Psi'(\vec{q_0}, \vec{r}) \cdot u(\vec{q_0}, \vec{r}) \end{aligned} \qquad (4.2)$$

where $W(\vec{q_0}, \vec{r})$ is the weighting function and L is the diameter of the sphere. $W(\vec{q_0}, \vec{r})$ was chosen to be a Gaussian with a phonon amplitude of a $1/e$ at the boundary of the crystal. However, there was no physical justification for this assumption. To calculate the effect of confinement on the Raman spectrum, Ψ' is expanded in a Fourier series as:

$$\Psi'(\vec{q_0}, \vec{r}) = \int d^3q \, C(\vec{q_0}, \vec{q}) \exp(\vec{q} \cdot \vec{r}). \qquad (4.3)$$

The Fourier coefficient $C(\vec{q_0}, \vec{q})$ is determined as:

4.1 The Phonon confinement: RCF-model

$$C(\vec{q_0}, \vec{q}) = \frac{1}{(2\pi)^3} \int d^3r\, \Psi'(\vec{q_0}, \vec{r}) \exp(-i\vec{q} \cdot \vec{r}). \tag{4.4}$$

The phonon wavefunction is a superposition of eigenfunctions with wavevector $\vec{q}$, centered at $\vec{q_0}$. Assuming $\vec{q_0} = 0$ for one-phonon processes and neglecting the scaling factors, the Fourier coefficient for a spherical crystal is given by:

$$|C(0,q)|^2 \cong \exp-(q^2 L^2/4). \tag{4.5}$$

The first-order Raman spectrum $I(\omega)$ can then be written as:

$$I(\omega) = \int \frac{d^3q\, |C(0,\vec{q})|^2}{(\omega - \omega(\vec{q}))^2 + (\Gamma_0/2)^2}, \tag{4.6}$$

where $\omega(\vec{q})$ is the phonon dispersion and Γ_0 is the natural linewidth. It should be noted that in order to simplify the calculation, the Brillouin zone was assumed to be spherical and the dispersion curve isotropic. To adapt the Richter et al. model [38] to the nanowire shape and taking into account the fact that the phonon wavefunction should vanish at the boundary of the crystal, Campbell and Fauchet [39] examined various weighting function $W(\vec{q_0}, \vec{r})$ and found that the best weighting function, which agrees with the experimental data is a Gaussian with a value of $\exp(-4\pi^2)$ at the boundary of the crystal. The resulted Fourier coefficient and weighting function are $|C(0, q_1, q_2)|^2 \cong \exp(-\alpha q^2 L^2/16\pi^2)$ and $W(\vec{q_0}, \vec{r}) = \exp(-8\pi^2 r^2/L^2)$, respectively. For a cylinder with the length of L_2 and a diameter of L_1 the Fourier coefficients are calculated to be:

$$|C(0, q_1, q_2)|^2 \cong \exp\left(-\alpha q_1^2 L_1^2/16\pi^2\right) \cdot \exp\left(-\alpha q_2^2 L_2^2/16\pi^2\right) \tag{4.7}$$
$$\times \left|1 - \operatorname{erf}\left(\frac{iq_2 L_2}{\sqrt{32\pi}}\right)\right|^2.$$

The variable confinement parameter α is used to match the actual strength of the confinement with the available experimental data [39]. Since in the case of SiNWs the wires are not confined along the wire axis, d^3q in Equation 4.6 will be equal to qdq and

Chapter 4. Geometry dependence of the phonon modes in Si nanowires

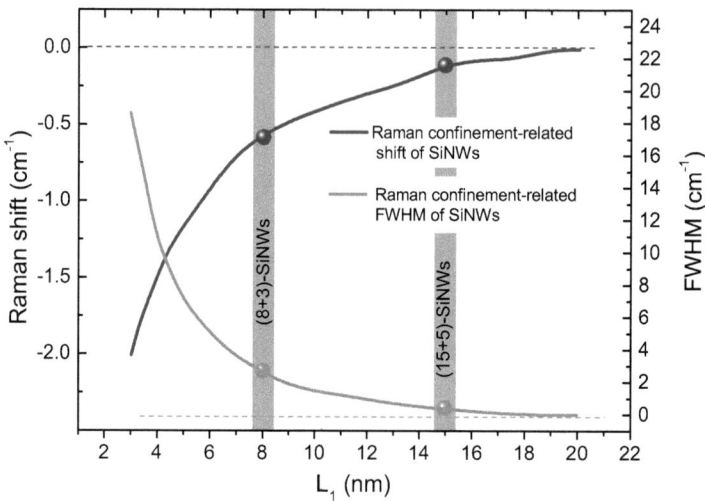

Figure 4.1: Calculated [29] confinement–related trends for Raman shift (blue solid line) and FWHM (green solid line) for SiNWs as a function of diameter (L_1). The blue (green) solid circles represent the pure confinement–related part of frequency shift (FWHM) of the SiNWs studied in this work. The diameter of the SiNWs was obtained from TEM measurements, presented in Chapter 2.3.

the L_2-terms in equation 4.7 can be neglected. To calculate the Raman spectrum of confined phonons, one integrates equation 4.6 with a fixed frequency ω over the whole Brillouin zone and then ω must be varied stepwise. For SiNWs, the first-order optical phonon dispersion can be approximated as $\omega(q) = [A + B\cos(q\pi/2)]^{0.5} + D$ [56], as reported by Piscanec et al. [29], and the Equation 4.6 can be modified as:

$$I(\omega) = \int \frac{qdq \exp\left(-q^2 L_1^2 / 16\pi^2\right)}{(\omega - ([A + B\cos(q\pi/2)]^{0.5} + D))^2 + (\Gamma_0/2)^2}. \tag{4.8}$$

A and B constants are $1.714 \times 10^{-5}\,\text{cm}^{-2}$ and $10^{-5}\,\text{cm}^{-2}$, respectively, which are obtained from neutron scattering experimental data for the TO-branch of Si [57].

4.1 The Phonon confinement: RCF-model

Table 4.1: The experimental data of Raman shift and FWHM compared with the pure confinement-related data based on RCF-model, obtained from Reference [29]. The Data for the bulk Si are obtained experimentally, using the Raman setup explained in Chapter 5.2.1.

Sample SiNWs	Diameter nm TEM	ω_0 cm^{-1} Raman	$\Delta\omega_0$ cm^{-1} Raman	$\Delta\omega_0$ cm^{-1} Ref. [29]	Γ_0 cm^{-1} Raman	$\Delta\Gamma_0$ cm^{-1} Raman	$\Delta\Gamma_0$ cm^{-1} Ref. [29]
(8+3)	8	519.1(4)	−0.9	−1	7	3	2.6
(60+5)	60	518.7(4)	−1.3	0	4	0	0
(60+20)	60	520.5(2)	+0.5	0	4	0	0
(15+5)	15	518.0(1)	−2	−0.1	7	3	0.3
Bulk Si	∞	520	0	0	4	0	0

Figure 4.1 shows the calculated trends in Raman shift and full width at half maximum (FWHM) for SiNWs as a function of diameter (data taken from Piscanec et al. [29]). In order to determine the confinement-related part of Raman shift and broadening of the Raman peaks in SiNWs samples studied in this work, SiNWs diameters reported in Chapter 2.3 were used. The confinement-related shifts and FWHMs are compared with the data obtained from Raman measurements (see Table 4.1).

As shown in Table 4.1, the experimental Raman data deviate, to some extent, from the theoretical values of the Raman shift in first-order optical mode and its FWHMs. Although the theory [29] predicts no confinement-related Raman shift, a shift of about −1.3 cm^{-1} and 0.5 cm^{-1} is observed for (60+5)-SiNWs and (60+20)-SiNWs, respectively. Moreover, for SiNWs with a diameter of 15 nm ((15+5)-SiNWs) the predicted confinement-related Raman shift and FWHM are −0.1 cm^{-1} and 0.3 cm^{-1}, respectively, while Raman measurements show a Raman shift of about −2 cm^{-1} and a FWHM of about 0.3 cm^{-1}.

In addition to confinement-related Raman shift and broadening, homogeneous and/or local heating of SiNWs resulted from excitation laser power, as well as tensile stress, can lead to redshift in optical Raman frequency. Compressive stress, on the other hand, results in a blueshift.

Chapter 4. Geometry dependence of the phonon modes in Si nanowires

The Raman frequencies and FWHM of SiNWs presented in Table 4.1 are the extrapolated data obtained from power dependent Raman measurements and therefore approximate the values at zero excitation power. Hence the redshifting related to the local heating by the excitation laser power is negligible. The effects of homogeneous and local heating of SiNWs on their Raman spectra are discussed in Chapters 6 and 7, respectively. Chapter 5 describes the tensile (compressive) stress–related red(bule)-shifting of Raman modes.

The important thing in science is not so much to obtain new facts as to discover new ways of thinking about them.

Sir William Lawrence Bragg

5

Pressure dependence of the vibrational properties of SiNWs

Recently, many groups have conducted theoretical and experimental studies of the confinement related effects on the elastic properties of nanostructured material. Concerning the semiconductor nanomaterials in view of future applications, only a few reports are available. Available data on SiNWs mechanical properties show significant differences [7–12]. Some reports show the decrease of the Young and the bulk moduli with respect to the bulk Si, depending on the diameter of SiNWs [2, 7–12, 58, 59], while others claim an opposite trend [2, 58, 59]. In this chapter, the pressure dependence of the vibrational properties of SiNWs is studied. Section 5.1 describes the theory of the anharmonicity in Raman spectra and the effect of the hydrostatic pressure on the vibrational properties of solids. In Section 5.2 the pressure dependence of three different SiNW samples are studied *via* Raman spectroscopy.

Chapter 5. Pressure dependence of the vibrational properties of SiNWs

The longitudinal and transversal optical (LTO) Raman shift and also the FWHM (full width at half maximum) (Section 5.2.2.4) of first optical Raman mode of SiNWs with diameter of $\sim 8\,\text{nm}$ under hydrostatic pressure are studied. Section 5.3 discusses the effect of the oxide shell on the vibrational properties of SiNWs.

5.1 Theory: Effect of pressure on the vibration properties of solids

One of the important research fields in nanomaterials, in view of future applications, is to determine their physical properties under mechanical stress. Raman spectroscopy has been extensively used for studying the structure of semiconducting materials and nanomaterials under variable pressure, and constitutes a reliable technique for monitoring the properties of materials under such conditions. Normally under pressure, the Raman peaks of semiconsucting materials shift towards higher energies (blue shift) and become broader (increase of FWHM)[60–63].

5.1.1 Pressure dependent force constant: Anharmonicity

The increase of Raman frequencies with the applied pressure corresponds to the changes in inter-nuclear spacing, which is coupled with the anharmonicities of inter-atomic interactions [63, 64]. Applying pressure changes the inter-atomic spacing in the crystal, which in turn affects the force constants. According to Sherman [64] and Wilkinson [65], applying pressure in the case of a harmonic approximation, where the force constant does not depend on the spacing between the lattice atoms in a solid, does not result in a Raman frequency shift. Anharmonicity results from the dependence of the force constant on the displacement. The potential function of crystal vibrations can be written as a sum of an attractive (V_A) and a repulsive (V_B) term:

5.1 Theory: Effect of pressure on the vibration properties of solids

$$V_A = -A - B(r_e + x)^{-n} \tag{5.1}$$

$$V_B = -C + D(r_e + x)^{-m} \tag{5.2}$$

where r_e is the equilibrium distance between two atoms, x is the displacement of the vibration and A, B, C and D are constants. Imposing an initial condition in the potential $V = V_A + V_B$, such as $(\partial V/\partial x)_{x=0} = 0$, and assuming the amplitude of vibrations (x) is small compared to the equilibrium distance between two atoms (r_e), the force constant k can be written as:

$$\frac{d^2V}{dx^2} = k = k_0 \left(1 - \frac{(n+m+3)x}{r_e}\right) \tag{5.3}$$

where

$$k_0 = m(m-n)Dr_e^{-(m+2)}. \tag{5.4}$$

The relative change in the force constant can be obtained by taking the derivative of Equation 5.3 as:

$$\frac{dk}{k_0} = -(n+m+3)\left(\frac{dr}{r_e}\right) \tag{5.5}$$

Integration of this equation results in the force constant as a function of the interatomic displacements:

$$k = k_0 r^{-(n+m+3)}. \tag{5.6}$$

With $dk/k_0 = 2d\omega/\omega_0$, for a three dimensional oscillation the Equation 5.5 can be written as:

$$\frac{d\omega}{dP} = -\left(\frac{\omega}{V}\right)\left(\frac{n+m+3}{6}\right)\frac{dV}{dP} \tag{5.7}$$

Chapter 5. Pressure dependence of the vibrational properties of SiNWs

where ω is the oscillating frequency and V is the occupied volume by the unit mass of the crystal.

The volume dependence of the vibration frequencies can be expressed with the Grüneisen parameter (γ), which is given as:

$$\gamma = -\frac{d\ln\omega}{d\ln V} = -\frac{V}{\omega}\frac{d\omega}{dV}. \tag{5.8}$$

In the case of the monoatomic crystals, the Grüneisen parameter can be calculated by replacing the values of V and ω with ar^3 and $bk^{1/2}$, respectively:

$$\gamma = -(\frac{r^3}{k^{1/2}})\frac{d(k^{1/2})}{d(r^3)} = -(\frac{r}{6k})\frac{dk}{dr}. \tag{5.9}$$

Comparing this equation with Equation 5.5 yields the Grüneisen parameter for monoatomic crystals as $\gamma = (n + m + 3)/6$. Thus, Equations 5.6 and 5.7 can be written for monoatomic crystals as:

$$k = k_0 r^{-6\gamma}. \tag{5.10}$$

and

$$\frac{d\omega}{dP} = -(\frac{\omega}{V})\gamma\frac{dV}{dP} \tag{5.11}$$

5.1.2 Elastic properties and Raman scattering

Stress (σ_{ij}) and strain (ϵ_{ij}) are generally symmetric second-rank tensors, which describe the force per unit area applied to an elementary solid cube and the resulting deformation induced in the crystal by atomic displacements, respectively. The strain in a medium is proportional to the applied stress as:

$$\epsilon_{ij} = \sum_{k,l} S_{ijkl}\sigma_{kl} \tag{5.12}$$

5.1 Theory: Effect of pressure on the vibration properties of solids

$$\sigma_{ij} = \sum_{k,l} C_{ijkl}\epsilon_{kl}, \qquad (5.13)$$

where the proportionality constants are the forth-rank compliance (S_{ijkl}) and stiffness (C_{ijkl}) tensors. Considering the symmetry of cubic crystals (*e.g.* silicon) the matrix of the elastic stiffness C_{ijkl} coefficients is expressed by:

$$C = \begin{pmatrix} C_{11} & C_{12} & C_{12} & 0 & 0 & 0 \\ C_{12} & C_{11} & C_{12} & 0 & 0 & 0 \\ C_{12} & C_{12} & C_{11} & 0 & 0 & 0 \\ 0 & 0 & 0 & \frac{1}{4}C_{44} & 0 & 0 \\ 0 & 0 & 0 & 0 & \frac{1}{4}C_{44} & 0 \\ 0 & 0 & 0 & 0 & 0 & \frac{1}{4}C_{44} \end{pmatrix} \qquad (5.14)$$

and compliance tensor (S_{ijkl}) is given by:

$$S = \begin{pmatrix} S_{11} & S_{12} & S_{12} & 0 & 0 & 0 \\ S_{12} & S_{11} & S_{12} & 0 & 0 & 0 \\ S_{12} & S_{12} & S_{11} & 0 & 0 & 0 \\ 0 & 0 & 0 & \frac{1}{4}S_{44} & 0 & 0 \\ 0 & 0 & 0 & 0 & \frac{1}{4}S_{44} & 0 \\ 0 & 0 & 0 & 0 & 0 & \frac{1}{4}S_{44} \end{pmatrix}. \qquad (5.15)$$

Volume isothermal compressibility χ is defined as the relative variation of volume per unit variation of pressure, and is commonly written as:

$$\chi = -\left(\frac{\partial \ln V}{\partial P}\right)_T, \qquad (5.16)$$

which in the case of Si can be expressed as $\chi = 3(S_{11}+2S_{12})$ and its reciprocal value is the bulk modulus $\chi^{-1} = B$ [63, 66]. Comparing Equation 5.7 with Equation 5.16 yields:

$$B^{-1} = \chi = -\frac{1}{\gamma}\frac{\partial(\ln \omega)}{\partial P}. \qquad (5.17)$$

Chapter 5. Pressure dependence of the vibrational properties of SiNWs

As mentioned in Appendix A.1 (Equation A.2) the bulk modulus B varies with the applied external pressure $(B(P) = B_0 + B'_0 P)$. After incorporating the pressure-dependent bulk modulus in Equation 5.17, the integration leads to:

$$\ln \omega = \int_{P_0}^{P} \frac{\gamma}{B_0 + B'_0 P} \, dP = \frac{\omega \gamma}{B'_0} \ln (B_0 + B'_0 P) \Big|_{P_0}^{P} \quad (5.18)$$

and

$$\omega(P) = \frac{\omega_0}{B_0^{\gamma \omega / B'_0}} (B_0 + B'_0 P)^{\frac{\gamma \omega}{B'_0}}. \quad (5.19)$$

Expanding this equation results in:

$$\omega(P) = \omega_0 + \omega_0 (\frac{\gamma}{B_0}) P + \omega_0 (\frac{\gamma^2}{2B_0^2} - \frac{\gamma B'_0}{2B_0^2}) P^2. \quad (5.20)$$

As it will be shown in Section 5.2, by comparing the Raman shift of SiNWs under pressure with Equation 5.20, valuable data such as the bulk modulus of SiNWs and its derivative can be exported.

5.1.3 Frequency shift of LTO-phonons under hydrostatic pressure

Stress or strain can affect the Raman frequencies. One of the first theoretical works concerning the effect of stress on the Raman modes was presented by Ganesan et al. [67]. According to their analysis, in diamond-like crystals the dynamic equation under pressure, which corresponds to the induced strain ϵ_{lm} has the form [67]:

$$m\ddot{u}_i = -(K^0_{ii} u_i + \sum_{klm} K^{(1)}_{iklm} \epsilon_{lm} u_k), \quad (5.21)$$

where

5.1 Theory: Effect of pressure on the vibration properties of solids

$$\frac{\partial K_{ik}}{\partial \epsilon_{lm}}\epsilon_{lm} = K^{(1)}_{iklm}\epsilon_{lm} \tag{5.22}$$

and m and u are the reduced mass of the two atoms and the relative displacement of the atoms in the unit cell, respectively. Due to the symmetry of silicon crystal, only three components of the $K^{(1)}$ must be considered. These are $K^{(1)}_{1111} = K^{(1)}_{2222} = K^{(1)}_{3333} = mp$, $K^{(1)}_{1122} = K^{(1)}_{2233} = K^{(1)}_{1133} = mq$ and $K^{(1)}_{1212} = K^{(1)}_{2323} = mr$ where p, q and r are the phenomenological deformation potential constants. These constants determine the change in the spring constants (K_{ik}) in different crystal direction, when a ϵ_{lm} strain is applied [67]. Using Equation 5.17 and taking into account the fact that the isothermal compressibility is $\chi = 3(S_{11} + 2S_{12})$, the Grüneisen-mode parameter can be expressed as:

$$\gamma = \frac{-[p + 2q]}{6\omega_0^2}. \tag{5.23}$$

Under hydrostatic pressure, the stress tensor can be written as:

$$\sigma = \begin{pmatrix} \sigma & 0 & 0 \\ 0 & \sigma & 0 \\ 0 & 0 & \sigma \end{pmatrix}. \tag{5.24}$$

The resulting strain is defined by Equations 5.12 and 5.15. Using Equations 5.24 and 5.12, the frequencies of the optical mode under hydrostatic pressure can be obtained by solving the following equation:

$$\begin{vmatrix} K_{11}\epsilon_1 + K_{12}(\epsilon_2 + \epsilon_3) - \lambda & K_{44}\epsilon_4 & K_{44}\epsilon_5 \\ K_{44}\epsilon_4 & K_{11}\epsilon_2 + K_{12}(\epsilon_1 + \epsilon_3) - \lambda & K_{44}\epsilon_6 \\ K_{44}\epsilon_5 & K_{44}\epsilon_5 & K_{11}\epsilon_3 + K_{12}(\epsilon_1 + \epsilon_2) - \lambda \end{vmatrix} = 0 \tag{5.25}$$

where $\lambda = \omega_0^2 - \omega^2$. The solution of the equation above yields the frequencies of the optical phonon mode under hydrostatic pressure:

Chapter 5. Pressure dependence of the vibrational properties of SiNWs

$$\omega(\sigma) = \omega_0 + \frac{\sigma\left[(S_{11} + 2S_{12})(p + 2q)\right]}{2\omega_0}. \quad (5.26)$$

Knowing the phonon frequencies of diamond like semiconductors under hydrostatic pressure (pressure dependence Raman measurements), this Equation can be used for determination of their elastic constants.

5.2 Raman scattering of silicon nanowires under hydrostatic pressure

Raman scattering experiments were performed in order to investigate the influence of hydrostatic pressure on the optical phonon mode (LTO) of SiNWs with a mean diameter of ~ 8 nm, as explained in Chapter 2.3.1. In this section, firstly the experimental Raman setup including the pressure cell will be described, followed by the results of the pressure dependent Raman spectra for SiNWs.

5.2.1 Experimental setup

5.2.1.1 Raman Spectrometer

The macro-Raman setup (see Figure 5.1) consists of a Dilor - XY 800 spectrometer equipped with a camera objective (Minolta MD 50 mm 1 : 1.7), and a triple monochromator. The first two monochromators are used as filters for the stray light rejection and for filtering out the laser. The corresponding gratings have 1800 groves per mm. The third stage has two interchangeable gratings with 1800 and 2400 groves per mm. The Raman signal is collected with a back-thinned LN_2 cooled charge coupled device (CCD) from Wright Instruments. The Raman spectra are excited with the $\lambda = 514.5$ nm line from an Ar^+ laser in the back-scattering geometry. The spectral resolution is about $1.0\,\text{cm}^{-1}$. Proper calibration of the spectra was achieved by recording the atomic spectral lines from a Neon lamp.

5.2 Raman scattering of silicon nanowires under hydrostatic pressure

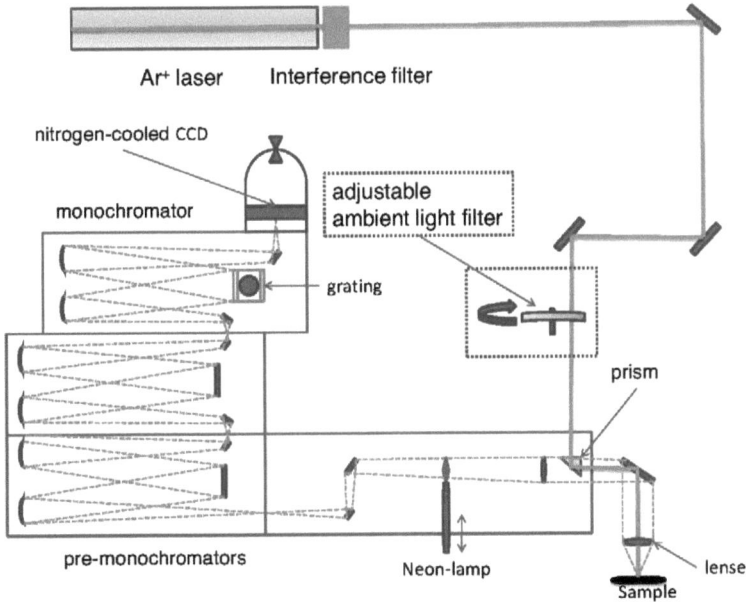

Figure 5.1: The schematic diagram of the macro-Raman setup used for the pressure dependence measurements.

5.2.1.2 Diamond anvil cell (DAC)

Measurements under hydrostatic pressure were conducted using the diamond anvil cell (DAC) technique, as described in detail in two review articles by Huber et al. [68, 69]. A gasket made of a specific nickel alloy (Inconel X-750) with a small hole in the middle was placed between two diamonds, which were pressed against each other using the Syassen-Holzapfel DAC as shown in Figure 5.2. The SiNWs were placed in the hole of the gasket together with a small ruby crystal and a pressure transmitting medium, consisting of a 4:1 mixture of methanol and ethanol, to ensure good hydrostatic conditions, at least for up to 10 GPa [70]. Under pressure, the gasket undergoes plastic deformation, thus enabling the hydrostatic compression of

Chapter 5. Pressure dependence of the vibrational properties of SiNWs

the samples between the diamonds. The ruby was used for the pressure calibration by measuring the shift of its R_1 line, which originates from a $d-d$ transition of the Cr^{3+} ions and provides a precise pressure Reference [70]. The spectra are fitted with Voigt functions after the background subtraction. There are some important issues related to the loading process of the DAC and pressure dependence measurements that should be considered, in order to have a successful and reliable outcome: the SiNW sample should be scraped from the substrate and placed in the middle of gasket hole (Figure 5.2). A very small piece of ruby crystal powder should be placed away from nanowires. A droplet of methanol and ethanol is added to the samples using a micropipette. The procedure should be carried out under the microscope to ensure that the sample is in the middle of the gasket, far from the ruby and neither of them has contact with the inner wall of the gasket. It is very important to wait about 20 min after modification of the pressure and to check the stability of the pressure in the gasket for each measurement.

5.2.2 Results and discussion

5.2.2.1 Is there any heating effect?

To guarantee that the measured shifts in the Raman frequencies are related to the pressure effect and not caused by the heating effect of SiNWs due to the excitation, the Raman frequencies of SiNWs were measured with various excitation powers. A high excitation power can increase the local temperature in the SiNWs, causing a red shift and a broadening of the Raman peaks [29–31]. Therefore, the laser power must be kept at low levels to avoid such local heating effects. This effect is negligible in bulk Si because of the better thermal conductivity of the bulk crystal [29, 71–74]. Here, the potential thermal effects on SiNWs are evaluated by studying the LTO mode as a function of the excitation power (Figure 5.3). Increasing the power, results in a reversible softening, with the rate of $1.27\,\text{cm}^{-1}/\text{mW}$, up to 13 mW. At excitation powers higher than 13 mW, the Raman shift saturates. This phenomenon will be discussed in detail in Chapter 7. The extrapolated zero-power Raman peak position is $518.9 \pm 0.6\,\text{cm}^{-1}$. The Raman red shift, caused by overheating, depends

5.2 Raman scattering of silicon nanowires under hydrostatic pressure

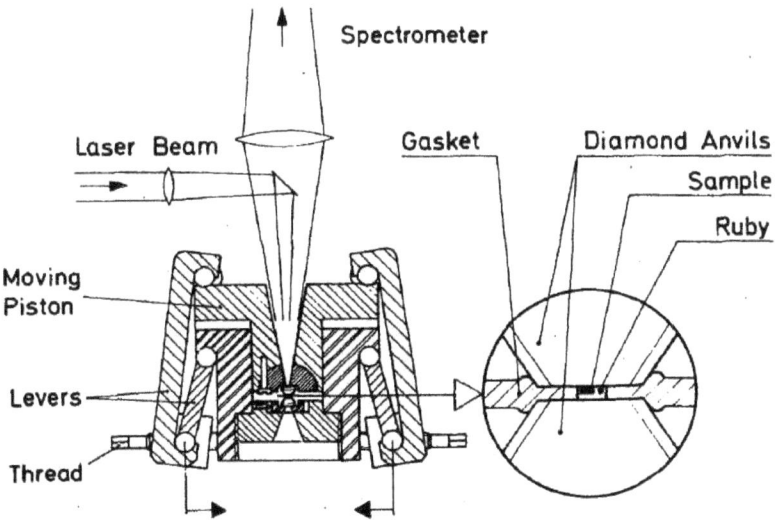

Figure 5.2: Schematic drawing of the cross-section of a Syassen-Holzapfel-type Diamond anvil cell and its enlarged view of the gasket. The space filled with the pressure transmitting medium has a diameter of 0.2 mm and a thickness of 0.1 mm [68].

on the thermal anchoring of SiNWs to the substrate [29] and the thermal conductivity of the surrounding gas [29, 30, 75]. These factors, together with the power density on the laser spot, which depends on the micro or macro experimental setup, determine the value of the frequency vs the power slope. This value has been reported to be $0.5\,\mathrm{cm}^{-1}/\mathrm{mW}$ [30] and $1\,\mathrm{cm}^{-1}/\mathrm{mW}$ [75, 76] for a macro Raman setup, and $\sim 2.5\,\mathrm{cm}^{-1}/\mathrm{mW}$ [77] for a micro Raman setup. Moreover, Reference [75] reported a linear dependence of this slope, with inverse thermal conductivity of the surrounding medium of SiNWs. For the high pressure setup used in this study, the medium is a methanol-ethanol mixture with a thermal conductivity of $0.204\,\mathrm{W/mK}$ for methanol [78] and $0.168\,\mathrm{W/mK}$ for ethanol [79], which are higher than that for the air ($0.024\,\mathrm{W/mK}$ [80]). Therefore, the laser induced overheating in the high pressure Raman measurements is about ten times lower than that in the air. Since

Chapter 5. Pressure dependence of the vibrational properties of SiNWs

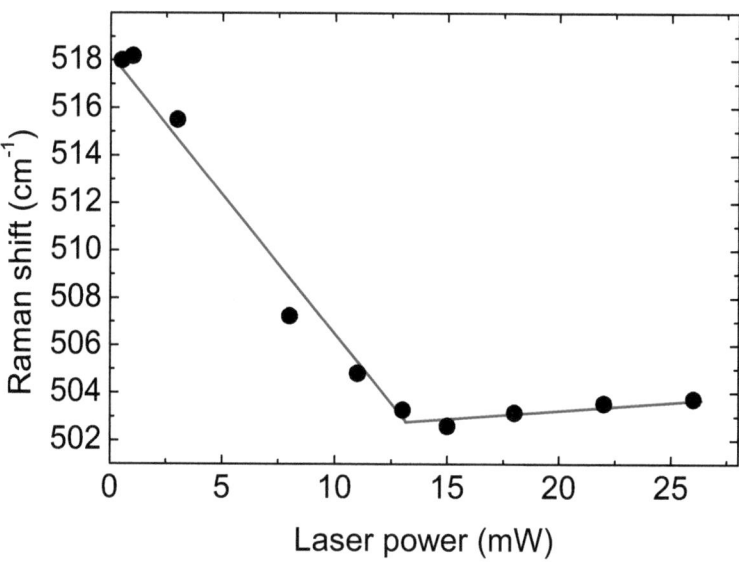

Figure 5.3: LTO frequency as a function of laser power, for the 514.5 nm excitation. The line is a guide to the eye.

the laser excitation power in this study was kept at about ∼ 1.5 mW, the overheating effect is negligible.

5.2.2.2 Raman shift under hydrostatic pressure

Figure 5.4 plots the measured Raman spectra of SiNWs as a function of pressure increases. One can observe an upshift and broadening of the LTO Raman peak, without any change of the lineshape. As mentioned in Chapter 2.3.1, a small group of the SiNWs present in the sample are nanochains. It can be assumed that the Raman signal collected from the necklace shaped nanostructures is negligible due to their low concentration (e.g. about 10% in (8+3)-SiNWs) and the small Raman cross-section of the crystalline part of the nanochains. The plots in Figure 5.5 exhibit the fitted

5.2 Raman scattering of silicon nanowires under hydrostatic pressure

pressure dependent change of the LTO peak. The effect of the compression and the decompression processes on the Raman spectra is reversible, as will be confirmed by the pressure dependence of FWHM of the LTO peak (see Section 5.2.2.4 and Figure 5.7).

As mentioned in Section 5.1.2 (Equation 5.20) the pressure dependence of the Raman frequencies of silicon-like crystals can be described by a quadratic function. The data in Figure 5.5 is fitted with such function:

$$\omega(P) = 519.11(4)\,[\text{cm}^{-1}] + 6.1(3)\,[\text{cm}^{-1}/\text{GPa}] \cdot P - 0.08(3)\,[\text{cm}^{-1}/\text{GPa}^2] \cdot P^2. \quad (5.27)$$

For comparison, Figure 5.5 also shows the room temperature pressure dependence for the bulk Si, taken from Reference [81]. The data are tabulated in Table 5.1. This shows that the pressure coefficient ($\frac{d\omega}{dP}$) of SiNWs (6.1(3) cm^{-1} GPa^{-1}) is about 17% higher than its bulk counterpart. The quadratic terms in the pressure dependent Raman frequencies ($-\frac{d^2\omega}{dP^2}$) originate from the nonlinear relationship between the relative lattice compression Δa and the external pressure P [82]. These in fact show a slight difference between the values for SiNWs and the bulk material. However, because the magnitude of the errors is compareble with the main values, no statement can be made. As discussed earlier in Section 5.1.3, the Raman frequencies ($\omega(P)$) of the optical modes under hydrostatic pressure (σ) are described by Equation 5.26. This can be written as:

$$\Delta\omega = \frac{(S_{11} + 2S_{12})(p + 2q)\sigma}{2\omega_0} = \frac{\chi(p + 2q)\sigma}{6\omega_0} \quad (5.28)$$

Using the elastic compliances of bulk silicon $S_{11} = 7.68 \times 10^{-3}$ GPa^{-1}, and $S_{12} = -2.14 \times 10^{-3}$ GPa^{-1}, and phenomenological deformation potential constants $p = -1.85/\omega_0^2$, $q = -2.31/\omega_0^2$ and $\omega_0 = 520$ cm^{-1}, the value of $\Delta\omega/\sigma = 5.046$ cm^{-1}/GPa is obtained from Equation 5.28. This value is in a good agreement with the experimental value of bulk pressure coefficient (5.2(3) cm^{-1}/GPa). For SiNWs, as shown in Table 5.1, this value is $\Delta\omega/\sigma = 6.1$ cm^{-1}/GPa. Assuming that: (A) the deformation potential constants p and q for SiNWs are the same as their bulk counterparts, i.e. $p_{\text{SiNWs}} = p_{\text{bulk}} = p$ and $q_{\text{SiNWs}} = q_{\text{bulk}} = q$, and (B) SiNWs are isotropic ma-

Chapter 5. Pressure dependence of the vibrational properties of SiNWs

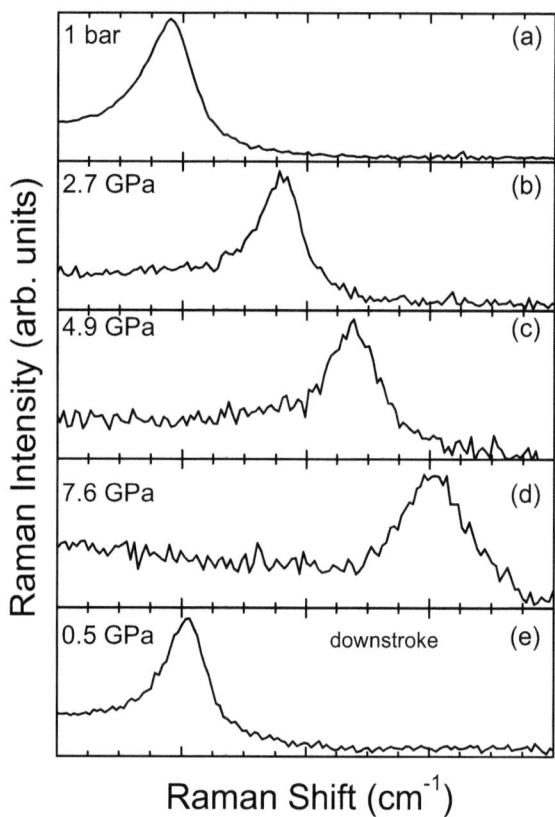

Figure 5.4: Raman spectra of (8+3)-SiNWs at various hydrostatic pressures, recorded upon pressure increase (a-d), and after pressure decrease (e). The excitation wavelength is $\lambda = 514.5$ nm and the excitation power is $P = 1.5$ mW.

5.2 Raman scattering of silicon nanowires under hydrostatic pressure

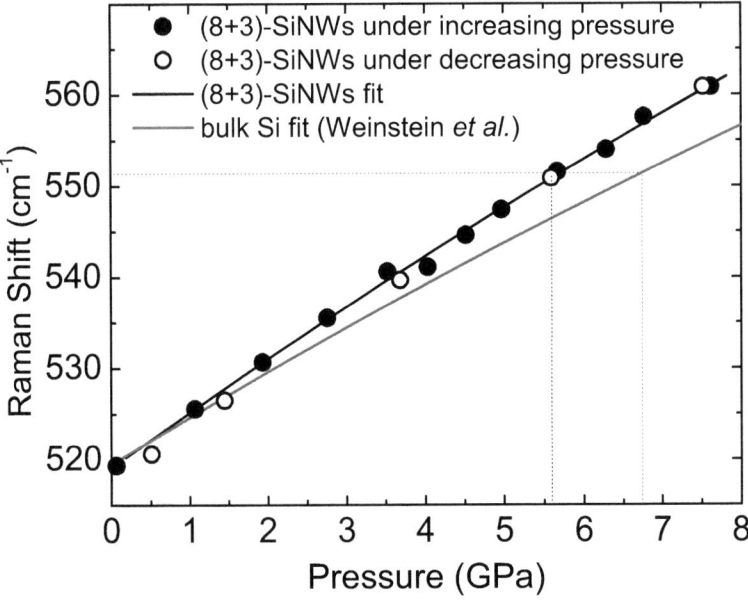

Figure 5.5: Pressure dependence of the LTO phonon in (8+3)-SiNWs. The black solid (open) circles denote data under increasing (decreasing) pressure. The red solid line represents the pressure dependence of bulk Si taken from Reference [81].

terials, *i.e.* $\chi = 3(S_{11} + 2S_{22})$, the compressibility ($\chi$) of SiNWs can be calculated as:

$$\chi_{\text{SiNWs}} = \chi_{\text{bulk}} \times \frac{\omega_{0_{\text{SiNWs}}}}{\omega_{0_{\text{bulk}}}} \times \frac{(\partial\omega/\partial P)_{\text{SiNWs}}}{(\partial\omega/\partial P)_{\text{bulk}}} \quad (5.29)$$

Using the well-known isothermal compressibility value of 10.5×10^{-3} GPa^{-1} for bulk silicon (Table 5.3), and ω_0 and $\partial\omega/\partial P$ values for SiNWs and bulk Si (Table 5.1), the isothermal compressibility of SiNWs can be calculated as $\chi = 12.0(6) \times 10^{-3}$GPa^{-1}. The bulk modulus of (8+3)-SiNWs is the inverse of the isothermal compressibil-

Chapter 5. Pressure dependence of the vibrational properties of SiNWs

Table 5.1: Phonon frequency at zero pressure, linear- and quadratic-pressure coefficients, γ/B_0 and $\gamma\chi$ parameters for (8+3)-SiNWs. The corresponding values for bulk Si from References [81] and [85] are also included.

Sample	ω_0	$\frac{d\omega}{dP}$	$-\frac{d^2\omega}{dP^2}$	$\frac{\gamma}{B_0}$	$(\frac{\gamma^2}{2B_0^2} - \frac{\gamma B_0'}{2B_0^2})$	Ref.
			($\times 10^{-2}$)	($\times 10^{-3}$)	($\times 10^{-4}$)	
	cm^{-1}	(cm GPa)$^{-1}$	(cm GPa)$^{-2}$	GPa^{-1}	GPa^{-2}	
(8+3)-SiNWs	519.11(4)	6.1(3)	8(3)	11.7(8)	1.5(5)	this study
Bulk Si	519.5(8)	5.2(3)	7(2)	10.0(6)	1.3(4)	[81]
Bulk Si	518.6	5.5	8.6	10.61	1.658	[85]

ity χ (see Section 5.1.2), and is found to be $B_0 = 83(4)$, which is about 17(3)% lower than that of the bulk silicon. The Grüneisen parameter can be obtained from Equation 5.23. The calculated Grüneisen parameter for (8+3)-SiNWs ($\gamma = 1.08(3)$) is presented in Table 5.3. This value shows no significant deviation from its bulk counterpart, due to the assumption that the phenomenological deformation potential constants remain unchanged. The quadratic terms in the pressure dependence of the Raman frequencies, as calculated in Section 5.1.2 (Equation 5.20), depend on B_0, as well as B_0'. This quadratic term for the bulk silicon is equal to $7(2) \times 10^{-2}$. Inserting the values of the bulk modulus, and the Grüneisen parameter for the bulk silicon in this equation, yields the first pressure derivative of the bulk modulus $B_0' = 4.0(7)$, which is in good agreement with the values for bulk silicon reported by Hu et al. ($B_0' = 4.24$ [83] ; $B_0' = 4.16$ [84]). Using the approach mentioned above, the B_0' was calculated for (8+3)-SiNWs as $B_0' = 3.0(8)$, which is about 25% lower than the calculated value for the bulk silicon ($B_0' = 4.0(7)$).

5.2.2.3 Relation between Bulk modulus and lattice constant of SiNWs

Determination of crystal structures, lattice constants, bulk moduli, and other static and dynamic properties of solids is possible using pseudopotential and total energy methods [86]. The only inputs in these methods are the atomic number and the mass of the atoms in the unit cell. Nevertheless, due to the complexity of the asso-

5.2 Raman scattering of silicon nanowires under hydrostatic pressure

ciated calculations, some empirical methods are developed to allow the researchers to estimate the material properties in a simplified manner. Reference [87] suggested an empirical formula to determine the bulk modulus of diamond and zinc-blend semiconductors. The formula results from the role of covalency in determining the bulk moduli of these semiconductor materials:

$$B_0(\text{GPa}) = (c - 220\lambda) \cdot d(\text{Å})^{-3.5}, \tag{5.30}$$

where B_0 is the bulk modulus, d is the nearest-neighbor distance, and λ is the ionicity empirical parameter ($\lambda = 0, 1$, and 2 for IV, III-V, and II-VI semiconductors, respectively). Reference [87] reported the proportionality constant c (Equation 5.30) to be 1971 for bulk moduli of SiNWs and the bulk Si. Table 5.2 shows a list of experimentally obtained lattice parameters and the bulk moduli of the bulk silicon. By incorporating these values in Equation 5.30, the average value of c is estimated to be $\sim 1962\,\text{GPa}\text{Å}^{3.5}$.

In a more recent work [88] Lam *et al.* introduced an analytic relation between the pressure derivative of the bulk modulus and the lattice constant:

Table 5.2: Lattice parameter a, nearest neighbor distance d and bulk moduli of SiNWs are compared to their counterparts in bulk Si.

Sample	a Å	$d^{-3.5}$ Å	B_0 GPa	Reference
(8+3)-SiNWs	5.57(5)	0.046(1)	–	this study
(60+5)-SiNWs	5.44(5)	0.049(1)	–	this study
SiNWs (15 nm)	5.437	0.0499	–	[51]
SiNWs (15 nm)	5.435	0.0500	–	[14]
SiNWs (70 nm)	5.423(2)	0.0504(1)	123(5)	[53]
SiNWs (22 nm)	5.448	0.0496	–	[52]
Bulk Si	5.43	0.05017	100(2)	[50]
Bulk Si	5.435	0.0500	99.9	[81]
Bulk Si	5.435	0.0500	94.8	[85]

Chapter 5. Pressure dependence of the vibrational properties of SiNWs

$$B'_0(\text{GPa}) = \frac{10}{3} + \frac{10DR_0^2}{9\Omega_0 B_0}, \qquad (5.31)$$

where D is a material parameter, which can be obtained from the pseudopotential, and R_0 and Ω_0 are the equilibrium Wigner-Seitz radius and the equilibrium atomic volume, respectively.

For the bulk silicon with $D = 0.077\,\text{Ry}/(\text{a.u.})^2$, $R_0 = 3.18\,\text{a.u.}$, $\Omega_0 = 134\,(\text{a.u.})^3$, and $B_0 = 98\,\text{GPa} = 6.6667 \times 10^{-3}\,\text{Ry}/(\text{a.u.})^3$, the pressure derivative of the bulk modulus is calculated using Equation 5.31 as $B'_0 = 4.3$, which is consistent with the experimentally obtained value of 4.24 [83]. As mentioned in Chapter 2.3.1, the measured lattice parameter a for (8+3)-SiNWs is 5.57(5) Å, corresponding to nearest-neighbor distance of $d = a \times \sqrt{3}/4 = 2.41(2)$ Å. Inserting this value in Equation 5.30 and assuming $\lambda = 0$ for silicon yields the estimated bulk modulus of SiNWs as $B = 90(3)\,\text{GPa}$, which is $\sim 11\%$ lower than that of the bulk Si. Although this estimated value of the bulk modulus is not a perfect match with the calculated value of $B = 83(4)\,\text{GPa}$ obtained from the pressure dependent Raman measurements (Section 5.2.2.2), it is still in line with the trend. These results indicate a lower bulk modulus, thus a higher compressibility for (8+3)-SiNWs compared to the bulk Si. Reference [53] reported an increased bulk modulus (123±5 GPa) for SiNWs with $\sim 70\,\text{nm}$ diameter, as derived from the high pressure synchrotron measurements. However, their lattice parameter was 5.423 Å, which is $\sim 0.13\%$ lower than that of the bulk Si. Others reported a higher lattice parameter, consequently a lower bulk modulus [14, 51] (see Table 5.2). To make an estimation for the pressure derivative of SiNWs bulk modulus (B'_0), Equation 5.31 can be used. R_0 is defined as the radius of a sphere having the same volume as the volume per particle. For the diamond-like structure this value can be calculated as a function of the lattice constant, using $V_{\text{WZ}} = a^3/8 = (4\pi/3)R_0^3$. Incorporating this in Equation 5.31, yields:

$$B'_0(\text{GPa}) = \frac{10}{3} + \frac{10D \left(\sqrt[3]{3/32\pi}\right)^2 a^2}{9\Omega_0 B_0} \qquad (5.32)$$

Using this equation, B'_0 for SiNWs can be obtained by substituting the values for the lattice parameter $a = 5.57$ Å, the bulk modulus $B_0 = 90(3)\,\text{GPa}$, and the equilibrium

5.2 Raman scattering of silicon nanowires under hydrostatic pressure

Table 5.3: Grüneisen parameter γ and isothermal volume compressibility χ calculated using the estimated bulk modulus from Equation 5.30. The corresponding values for bulk Si, derived from the data in References [81, 85] are also included for comparison.

Sample	a Å	$\chi \times 10^{-3}$ $(GPa)^{-1}$	B_0 GPa	B_0'	γ	B_0 (Equ. 5.30)	B_0' (Equ. 5.32)	Ref.
(8+3)SiNWs	5.57(5)	12.0(6)	83(4)	3.0(8)	1.08(3)	90(3)	4.44(4)	-
Bulk Si	5.435	10.0	99.9	4.11	0.98(6)	100	4.3	[81]
Bulk Si	5.435	10.5	94.8	2.5	1.08	—	-	[85]

atomic volume $\Omega_0 = 134\,(\text{a.u.})^3$ of SiNWs in Equation 5.32. The result is $B_0' = 4.44(4)$, which differs from the value of $3.0(8)\,\text{GPa}$ determined in Section 5.2.2.2.

5.2.2.4 Raman linewidth of SiNWs under hydrostatic pressure

Broadening of the Raman lines under high pressure is a well-known phenomenon in crystals. The anharmonic decay of the phonons into those with lower frequencies is an important relaxation process in semiconductors. The pressure dependence of the Raman linewidths is due to the influence of the phonon dispersion on the final states of the decay, which are affected by the pressure dependence of the phonon frequencies. For silicon, the FWHM increases linearly with the pressure as $\Gamma_{2Tg}(P) = 1.08 + 0.137 P$ [11]. According to the energy and momentum conservation, the LTO phonon can decay into two phonons with opposite momentum and the total energy equal to the sum of the primary LTO-phonon energies. There are some possibilities (decay channels) regarding the final states of the decay. In the case of Si with no applied pressure, the possible decay channels are LTO→LA+TA and LTO→LA+LA [1]. The decay into one optical and one acoustic phonon is kinematically forbidden. Previously, Debernardi et al. [89] calculated the anharmonic phonon lifetimes of Si using the Density Functional Perturbation theory. The lifetime of the LTO mode at the zone center can be written as [90]:

---

[1] LA: longitudinal acoustic phonon, TA: transversal acoustic phonon

Chapter 5. Pressure dependence of the vibrational properties of SiNWs

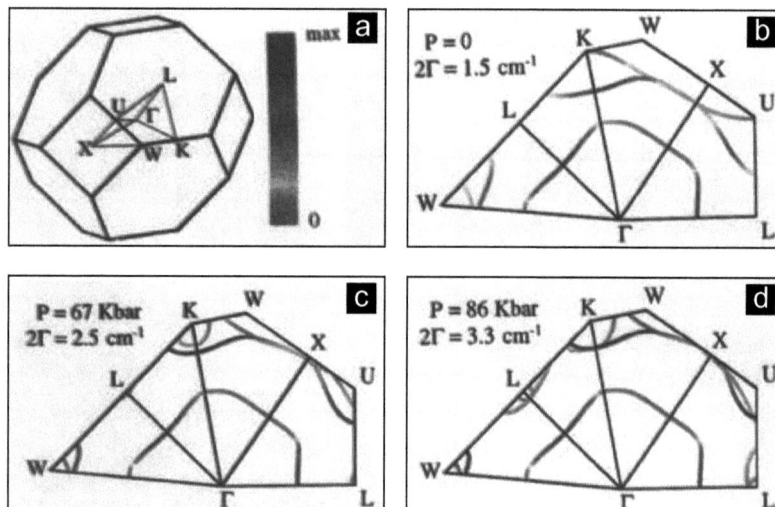

Figure 5.6: Wavevector resolved final state spectra of silicon at different pressures, taken from Reference [11]. (a) Brillouin zone, (b), (c), and (d) are the maps at 0 GPa, 6.7 GPa, and 8.6 GPa, respectively.

$$\Gamma = \frac{\pi \hbar}{16 N^3 M^3 \omega_{\text{LTO}}(0)} \sum_{q,j_1,j_2} \left[\left(\frac{\partial^3 E}{\partial u_{\text{LTO}}(0) \partial u_{j_1}(q) \partial u_{j_1}(-q)} \right)^2 \right.$$

$$\left. \times \frac{n_{j_1}(q) + n_{j_2}(-q) + 1}{\omega_{j_1}(q) \omega_{j_2}(-q)} \times \delta\left(\omega_{\text{LTO}}(0) - \omega_{j_1}(q) - \omega_{j_2}(-q)\right) \right],$$
(5.33)

where E is the crystal's energy, N is the number of the unit cells in the crystal, n is the thermal occupation number of the phonons, M is the atomic mass, j indicates the phonon branches, and $u_j(q)$ is the amplitude of j-phonon-mode with a wave vector of q.

Figure 5.6 shows the wavevector resolved final state spectra of Si, *i.e.* the probability per unit time that the LTO-phonon decays into one mode with a wavevector q. The

5.2 Raman scattering of silicon nanowires under hydrostatic pressure

q dependent part of Equation 5.33 expresses this probability within the summation term. As shown in this equation, the δ-function is only unequal zero when the energy remains consistent *i.e.* the energy of the LTO-phonon is equal to the sum of the energies of the generated phonons. The only non zero regions of the Brillouin zone (BZ), can be expressed by a 3D surface. The cross-section of this surface with high symmetry planes of the BZ is shown in Figure 5.6. Figure 5.6(a) shows the sketch of the BZ, with the color scale expressing the magnitude of the decay function (from red-low to violet-high). At zero pressure (Figure 5.6(a)) approximately at the middle of the BZ a curve is shown, which corresponds to the Klemens decay channel[2]. On this curve the energy of the LO-phonon reaches the half of the LTO-phonon's energy ($2\omega_{LA} = \omega_{LTO}$). As shown previously in Chapter 3.1 (Figure 3.1, dispersion relation of silicon) in the middle of the BZ the LA-branch-energy reaches the half of the energy of the LTO-branch. This situation enables the decay of LTO→LA+LA. In addition, the branches near the zone's edge in X, W, and K directions show the existence of final decay channels. This branches are related to the case, where $\omega_{LTO} = \omega_{TA} + \omega_{LA}$. In L and K directions the energy of the TA branches is very low (see Figure 3.1) so that the summation of this energy and the energy of the LA branch do not suffice to reach the energy of the LTO-mode. Applying pressure on Si causes an increase in the LTO modes energies as well as a decrease in the TA mode energies [81]. This effect enables a new decay channel, which in ambient conditions is forbidden. As shown in Figure 5.6 (c) and (d), applying pressure enables new decay possibilities in L and K directions, which are the channels with the LO and TA modes as the final decay states (LTO→LO+TA). Activation of a new channel most certainly results in a stepwise decrease (increase) of the lifetime (FWHM) of the LTO phonons.

From the measured Raman spectra of SiNWs the FWHM of the LTO peak can be extracted. The observed lineshape corresponds to the convolution of a Lorentzian peak with the Gaussian profile, *i.e.* a Voigt profile [91]. The Lorenzian lineshape corresponds to the Raman peak, and the Gaussian profile to the instrumental profile. To determine the instrumental Gaussian profile the FWHM of neon lines was used

---

[2] The Klemens decay channel is defined as the decay of LTO mode into two acoustic phonons, which belong to the same branch (both transversal or both longitudinal)

Chapter 5. Pressure dependence of the vibrational properties of SiNWs

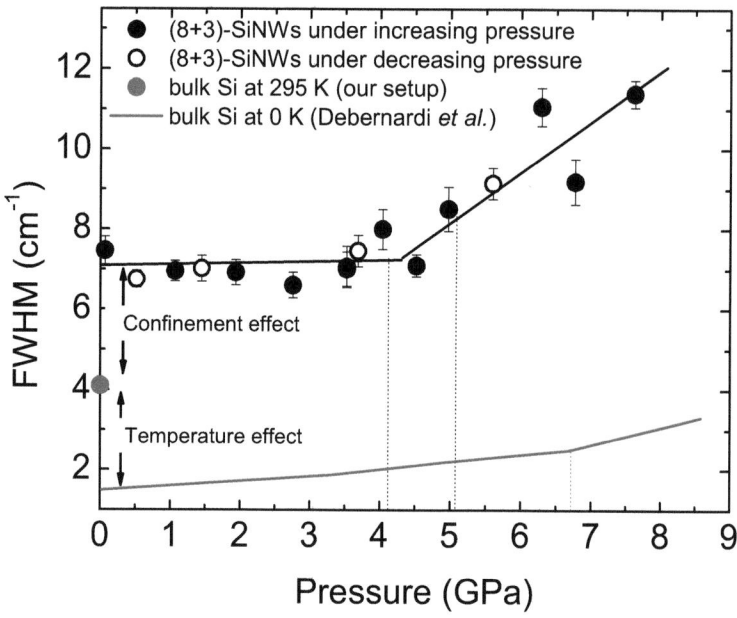

Figure 5.7: Pressure dependence of the LTO mode FWHM in (8+3)-SiNWs and bulk Si. The filled (open) circles refer to increasing (decreasing) pressure for SiNWs. The red solid line corresponds to the calculated trend for bulk Si at 0 K from Reference [89]. The red solid circle is our measured FWHM of bulk Si at 295 K.

(2.6 cm^{-1}). Using this fixed FWHM the SiNWs Raman peaks were deconvoluted.

The pressure dependence of the LTO peak FWHM for SiNWs is shown in Figure 5.7. The filled (open) circles refer to pressure increase (decrease). The trends are fully reversible. The plots in Figure 5.7 also show the pressure dependence of the bulk Si FWHM at 0 K (red solid line), which are the calculated data taken from Reference [89]. The FWHM of SiNWs at ambient pressure is ~ 7 cm^{-1}, which is ~ 6 cm^{-1} higher than the theoretical value for the bulk Si (1.4 cm^{-1} at 0 K). Figure 5.7 also shows the FWHM of the bulk Si measured in ambient conditions (red solid circle)

5.2 Raman scattering of silicon nanowires under hydrostatic pressure

using the Raman setup explained in Section 5.2.1. According to Reference [92], the bulk Si Raman FWHM changes from $1.4\,\text{cm}^{-1}$ (theoretical value) to $4\,\text{cm}^{-1}$ for a temperature increase from 0 to 295 K (see Chapter 6). It can be concluded that the $2.5\,\text{cm}^{-1}$ residual FWHM difference between (8+3)-SiNWs and the calculated FWHM is related to the ambient temperature during measurements (increase of the FWHM from the calculated value at about 0 K to the corresponding value at ambient temperature). The remainder of this difference (from 4 to $7\,\text{cm}^{-1}$) can be attributed to the phonon confinement, according to Reference [29], where a confinement related broadening of $\sim 2.5\,\text{cm}^{-1}$ was reported for SiNWs with a 8 nm diameter (see discussion in Chapter 4). As shown in Figure 5.7, for pressures of up to 7 GPa, the FWHM increase of the bulk silicon is attributed to the decay of LTO into LA + TA (94% probability) and LA + LA (6% probablity) phonons [89]. Above this critical pressure ($\sim 7\,\text{GPa}$), a new channel (LTO→LO + TA) begins to contribute [89], causing an increase in the slope of the FWHM-pressure diagram. In the case of SiNWs, this new decay channel becomes active at $\sim 4.5\,\text{GPa}$, causing a steep incline in FWHM (Figure 5.7). As mentioned before, the activation of the LTO→LO + TA decay channel is resulted from the fact that under pressure the energy of the LTO branch increases and that of the TA mode decreases. The activation of this channel at lower pressure ($\sim 4.5\,\text{GPa}$) means that the pressure-induced shift of the phonon energies in the dispersion relation of SiNWs must be more enhanced compared to their bulk counterpart. This statement is further validated based on the knowledge that the LTO phonons of SiNWs have a 17% higher pressure coefficient. It should be stressed that the estimatimation of the pressure for the activation of LO + TA channel for SiNWs requires the pressure coefficient of TA mode, which is unfourtunelly not determined for (8+3)-SiNWs. Nevertheless, as a rough estimate, the bulk Si Raman position at 7 GPa ($551\,\text{cm}^{-1}$) was compared with the pressure required to generate the same Raman shift in SiNWs ($\sim 5.5\,\text{GPa}$). The difference of 1.5 GPa is of the same magnitude as the decreased decay onset.

Chapter 5. Pressure dependence of the vibrational properties of SiNWs

5.2.3 Conclusions

The effect of pressure (up to 8 GPa) on the LTO mode of (8+3)-SiNWs (diameter of ~ 8 nm) was studied by Raman spectroscopy, and a more pronounced pressure dependence was detected compared to the bulk Si. By combining the resulted data from Raman spectra of (8+3)-SiNWs under hydrostatic pressure with the lattice anharmonic dynamical theory in diamond-like semiconductors, the bulk modulus B_0 and its pressure derivative B' for SiNWs were determined. The estimated B_0 and B' for SiNWs were 17(3)% and 25% lower than their bulk counterparts, respectively. Using a phenomenological formula [87, 88] and the lattice parameter extracted from the TEM measurements, the bulk modulus and the Grüneisen parameter of SiNWs were also estimated. Moreover, it was found that the remarkable FWHM increase, which was attributed to the pressure induced activation of LTO→LO + TA decay channel for (8+3)-SiNWs is achievable at a lower pressure (~ 4.5 GPa) compared to the bulk silicon (~ 7 GPa).

5.3 The effect of oxide layers on the vibrational properties of silicon nanowires

In this section high pressure Raman experiments performed on two SiNW samples with different oxide layers are presented. The pressure dependence of Raman spectra of these two samples were recorded, in order to examine the effect of the thickness of SiO_x layer on the vibrational properties of SiNWs under pressure. The samples were (60+5)-SiNWs and (60+20)-SiNWs, which possess thin and thick oxide layers, respectively. These two samples were described in Chapter 2.3.2. As discussed in Chapter 4, for SiNWs with a diameter above 20 nm no confinement-related Raman shift or broadening is to be expected. Moreover, because the experimental setup in this section is the same as that described in Section 5.2.2.2, a Raman shift or a broadening resulting from local heating by the laser excitation power can be neglected.

In Figure 5.8, two Raman spectra at room temperature, which correspond to the

5.3 The effect of oxide layers on the vibrational properties of silicon nanowires

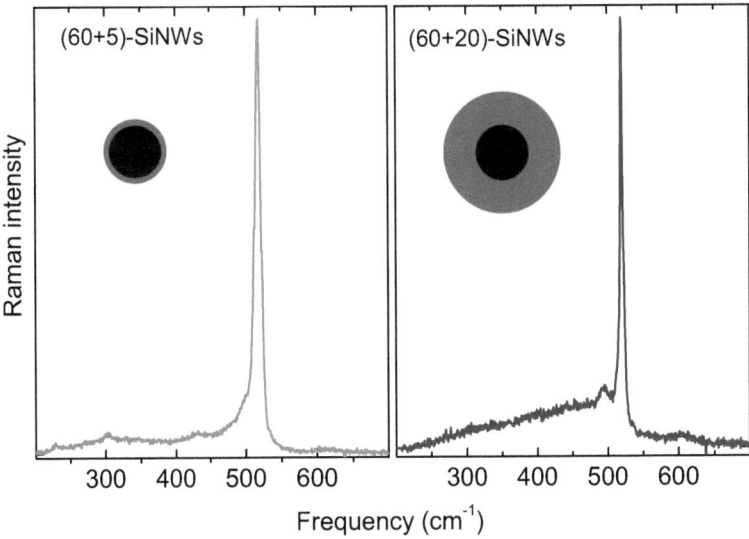

Figure 5.8: Raman spectra of (60+5)-SiNWs and (60+20)-SiNWs. The broad profile and the peak at about 495 cm^{-1} in (60+20)-SiNWs are characteristic Raman features of SiO$_2$, and SiO$_x$, respectively. Since these features are more prominent in (60+20)-SiNWs sample, one can conclude that the (60+20)-SiNWs has a thicker oxide layer than the (60+5)-SiNWs sample.

(60+5)- and (60+20)-SiNW samples, are presented. Next to the prominent first-order optical Raman peak of Si at about 520 cm^{-1}, a weak broad background related to the amorphous SiO$_2$ and the second-order feature of Si at 300 cm^{-1} can be observed for the (60+5)-SiNW sample. Aside from these peaks, the Raman spectra of the (60+20)-SiNWs show a relatively intense broad structure between 300 – 500 cm^{-1}, which is commonly attributed to the SiO$_2$ shell surrounding the SiNWs [93]. The Raman signal arising from the background features is more pronounced in the case of the (60+20)-SiNWs compared to the (60+5)-SiNWs. Moreover, the peak at about 495 cm^{-1}, which has been assigned to the SiO$_x$ [94], is more prominent in the case of the (60+20)-SiNWs. These characteristic confirm the

Chapter 5. Pressure dependence of the vibrational properties of SiNWs

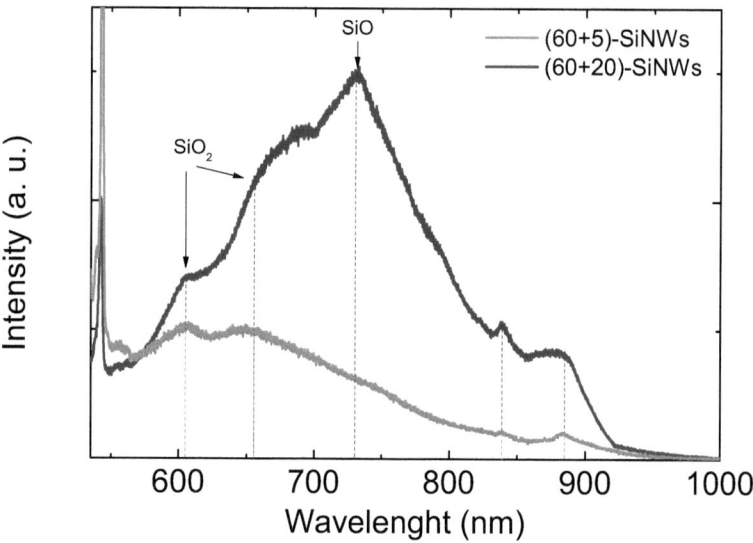

Figure 5.9: PL spectra of (60+5)-SiNWs and (60+20)-SiNWs. The PL spectrum of (60+20)-SiNWs shows beside SiO_2 broad peaks at about 600 nm and 700 nm an intense peak attributed to the SiO at about 740 nm [95].

higher thickness of the SiO_x shell for the (60+20)-SiNWs.

Figure 5.9 plots the Photoluminescence spectra of (60+5)- and (60+20)-SiNWs. The spectrum of (60+5)-SiNWs shows the characteristic broad peaks of SiO_2 at about 600 nm and 700 nm. In the case of (60+20)-SiNWs beside the characteristic peaks for SiO_2, a intense peak at about 740 nm was observed. This peak is attributed to the SiO [95]. This means, that the oxide layer of (60+20)-SiNWs consisted not only of amorphous SiO_2, but also SiO.

5.3 The effect of oxide layers on the vibrational properties of silicon nanowires

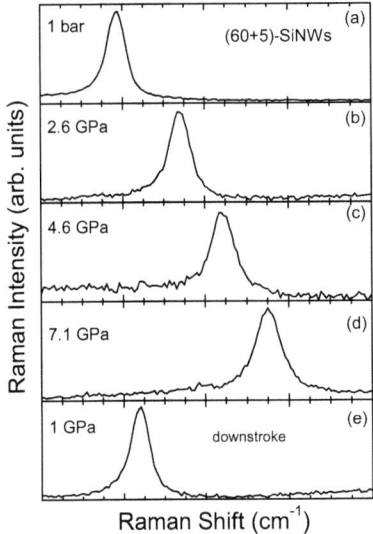
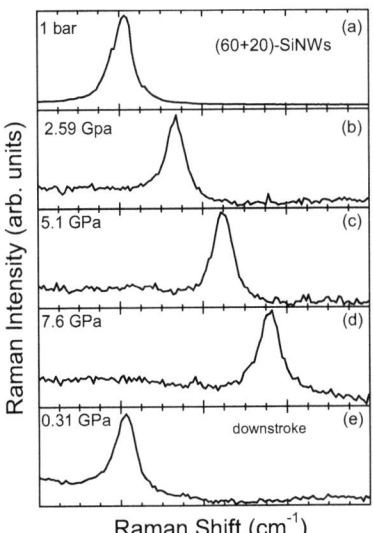

Figure 5.10: Raman spectra of (60+5)-SiNWs under ambient conditions (a), high pressure (b – d) and after decreasing the pressure to 1 GPa (e), measured with 514.5 nm laser line excitation.

Figure 5.11: Raman spectra of (60+20)-SiNWs under ambient conditions (a), high pressure (b – d) and after decreasing the pressure to 0.31 GPa (e), measured with 514.5 nm laser line excitation.

5.3.0.1 Raman shift of (60+5)- and (60+20)-SiNWs under pressure

The Raman spectra of the (60+5)-SiNWs and (60+20)-SiNWs, recorded under different hydrostatic pressures using 514.5 nm excitation laser line, are shown in Figures 5.10 and 5.11, respectively. Both figures show the optical Raman mode of SiNWs in ambient conditions (a), under various hydrostatic pressures (b – d) and finally, under pressure release (e). Upon pressure application, the observed Si Raman peak shifts towards higher energies for both samples. Also, no distinctive pressure-related changes were observed in the spectral line shape.

Chapter 5. Pressure dependence of the vibrational properties of SiNWs

The dependence of the optical Raman mode position in both samples, on increasing (black solid circles) and decreasing pressure (open circles), are illustrated in Figures 5.12 and 5.13, respectively. The experimental results for increasing pressure were fitted with a quadratic least-squares fit, as discussed previously in Section 5.1.2 (Equation 5.20), as:

$$\omega_{(60+5)\text{-SiNWs}}(P) = 518.7(4) + 6.4(3) \cdot P - 18(4) \times 10^{-2} \cdot P^2 \quad (5.34)$$
$$\omega_{(60+20)\text{-SiNWs}}(P) = 520.5(2) + 5.1(1) \cdot P - 6(2) \times 10^{-2} \cdot P^2 \quad (5.35)$$

where $\omega(P)$ is the Raman shift as a function of pressure. The fit parameters for (60+5)-SiNWs and (60+20)-SiNWs are summarized and compared with those of the bulk Si in Table 5.4. Upon pressure release, the frequency shifts are fully reversible, exhibiting no apparent hysteresis and the original SiNW spectrum is restored at 1 bar. For comparison, the pressure dependence of the optical Raman mode for bulk Si at room temperature are presented in Figures 5.12 and 5.13 (red solid line taken from Weinstein et al. [81], see Table 5.4).

As listed in Table 5.4, the pressure coefficient $(d\omega/dP)$ of (60+5)-SiNWs (6.4(3) cm^{-1}/GPa) differs from that of (60+20)-SiNWs (5.1(1) cm^{-1}/GPa) and the bulk silicon (5.2(3) cm^{-1}/GPa). Moreover, the Raman frequency at zero pressure (ω_0) for (60+5)-SiNWs (518.7(4) cm^{-1}) varies from that of (60+20)-SiNWs (520.5(2) cm^{-1}) and bulk silicon (519.5(8) cm^{-1}). The Raman frequency discrepancy of (60+5)-SiNWs ((60+20)-SiNWs) at zero pressure indicates a slight lattice expansion (contraction) compared with that of the bulk silicon [96], which will be discussed later. The quadratic pressure coefficient of (60+5)-SiNWs ($-18(4) \times 10^{-2}$ (GPa cm)$^{-2}$), in Table 5.4, is more pronounced compared with those of (60+20)-SiNWs ($-6(2) \times 10^{-2}$(GPa cm)$^{-2}$) and bulk silicon ($-7(2) \times 10^{-2}$(GPa cm)$^{-2}$). This result will be discussed in detail later. As discussed in Section 5.2.2.2, for the bulk silicon $\Delta\omega/\sigma$ equals 5.046 cm^{-1}/GPa. Comparing this value with the pressure coefficients of 6.4±0.3 cm^{-1}GPa^{-1} and 5.1±0.1 cm^{-1}/GPa for (60+5)-SiNWs and (60+20)-SiNWs, respectively, the corresponding $\chi(p+2q)$ terms in Equation 5.28 become 12.8 and 10.2, which are about 27% higher and 4% lower than that for the bulk silicon. Under the assumption that the deformation potential constants remain unchanged (Section 5.2.2.2) for (60+5)- and (60+20)-SiNWs samples and using Equation 5.26, the

5.3 The effect of oxide layers on the vibrational properties of silicon nanowires

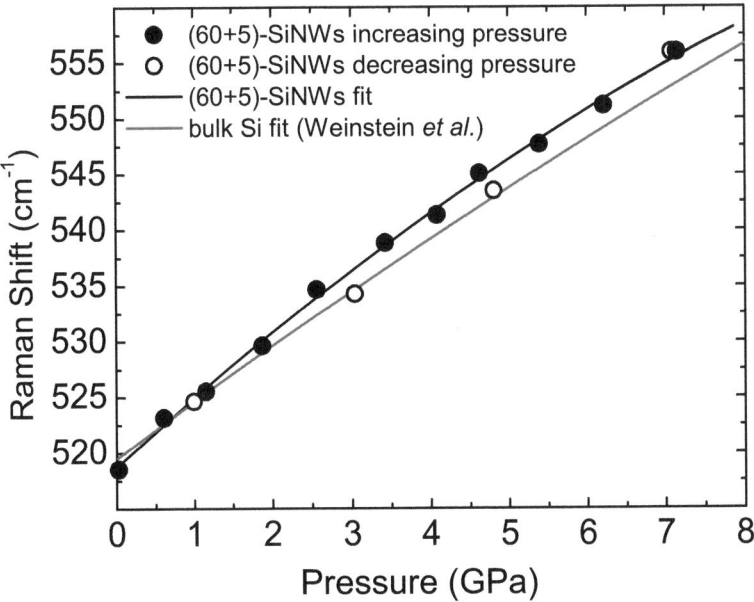

Figure 5.12: Pressure dependence of the optical mode in (60+5)-SiNWs. The filled (open) symbols denote data obtained for increasing (decreasing) pressure while the black solid line is the least square fitting. The red solid line corresponds to the pressure response of bulk Si [81].

isothermal compressibility (χ) and the bulk modulus ($B_0 = \chi^{-1}$) of SiNWs were calculated and listed in Table 5.4. The isothermal compressibility of (60+5)-SiNWs and (60+20)-SiNWs were found to be $\chi = 0.0125(9)$ GPa and $0.0100(6)$ GPa, corresponding to the bulk moduli $B_0 = 80(5)$ GPa and $100(5)$ GPa.

Using the fitting parameters in Table 5.4 and Equation 5.20 yield $\gamma/B_0 = 12.3(5) \times 10^{-2}$ GPa^{-1} for (60+5)-SiNWs and $9.8(2) \times 10^{-3}$ GPa^{-1} for (60+20)-SiNWs, as well as $\gamma\chi = 9.6(2) \times 10^{-3}$ GPa^{-1} for (60+5)-SiNWs and $8.7(1) \times 10^{-3}$ GPa^{-1} for (60+20)-SiNWs. Extracting the values for Grüneisen parameter γ requires the knowledge of the bulk modulus of the SiNWs, which was calculated before and listed in Table 5.4.

Chapter 5. Pressure dependence of the vibrational properties of SiNWs

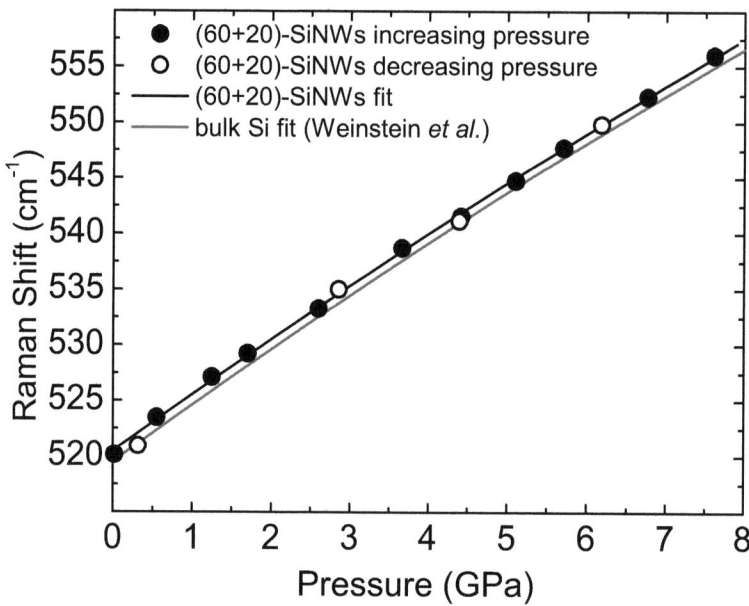

Figure 5.13: Pressure dependence of the optical mode in (60+20)-SiNWs. The solid (open) symbols denote the data obtained for increasing (decreasing) pressure while the black solid line is the least square fitting. The red solid line corresponds to the pressure response of bulk Si [81].

The values of Grüneisen parameter γ for (60+5)- and (60+20)-SiNWs are equal to their bulk counterpart (see Table 5.4). This is a consequence of the assumption that the deformation potential constants p and q for SiNWs are the same as those for the bulk silicon.

As mentioned before, the observed offset in the Raman frequency of (60+5)-SiNWs and (60+20)-SiNWs at zero pressure compared to the bulk Si indicates a slight lattice expansion and contraction, respectively. Many groups have found a larger lattice parameter for SiNWs compared to the bulk Si, using X-ray diffraction (XRD) [14, 97–99] and high resolution TEM [97, 100] techniques. This lattice expansion $(a - a_0)$

5.3 The effect of oxide layers on the vibrational properties of silicon nanowires

leads to a small stress, causing a Raman red shift ($\Delta\omega_0$), which can be calculated using $\Delta\omega = -n\gamma\omega_0(a - a_0)/a_0$ [101]. In this case n is the dimensionality of the materials ($n = 1$ for SiNWs), γ is the Grüneisen parameter (0.98) and ω_0 is the Raman frequency of the corresponding bulk material ($520\,\text{cm}^{-1}$). Inserting the Raman frequency shift for (60+5)-SiNWs at zero pressure ($\Delta\omega$) in the equation above and using the corresponding data from Table 5.4 yields a lattice parameter a of 5.445 Å, which is 0.3% higher than the value of its bulk counterpart. This value is consistent with the data obtained from the selected area electron diffraction (SAED) (inset of Figure 2.9(b)) and high resolution TEM (Figure 2.9(b)) images presented in Chapter 2.3.2.

According to Fukata et al. [98], a thin layer of amorphous SiO_x on the surface of SiNWs causes a shift of the Raman mode toward a lower wavenumber. This is observed in (60+5)-SiNWs. A thicker SiO_x layer, on the other hand, results in increased surface oxidation related compressive stress, which in turn causes the optical phonon peak to shift towards a higher wavenumber. This finding is consistent with the $\Delta\omega_0$ for (60+20)-SiNWs ($\omega_0 = 520.5\,\text{cm}^{-1}$), which predicts a smaller lattice parameter in comparison to its bulk counterpart (see Table 5.4). The efforts to measure the lattice parameter of (60+20)-SiNWs using high resolution TEM and SAED failed due to thick oxide layer as discussed in Chapter 2.3.2. The effect of the thin and the thick oxide layer on the pressure coefficient of SiNWs can be explained by the lattice expansion and the post oxidation lattice contraction discussed above. In a core-shell model for oxidized SiNWs under hydrostatic pressure, the bulk modulus of the SiO_x shell ($B_0 = 37.2(2)\,\text{GPa}$) [102] is lower than that of the crystalline silicon core. Thus, under hydrostatic pressure, the shell becomes more compressed than the silicon core due to its higher compressibility. Moreover, the pressure derivative of the bulk modulus for SiO_x ($B'_0 = 6.0(1)$) [102] is higher than the value for its bulk counterpart (4.24) [53]. The SiO_x surrounding the silicon core seems to be strengthened under pressure and accommodates a portion of applied external stress. Thus the pressure transferred to the core becomes smaller. Consequently, (60+20)-SiNWs become less compressed than (60+5)-SiNWs, which have thin layer of oxide.

Comparing the fit data of Equation 5.35 (quadratic part) with Equation 5.20, yields a 30% higher pressure derivative of the bulk modulus ($B' = 5.44$) for (60+5)-SiNWs

Chapter 5. Pressure dependence of the vibrational properties of SiNWs

Table 5.4: Phonon frequency at zero pressure (ω_0), phonon linear- ($\frac{d\omega}{dP}$) and quadratic-pressure coefficient ($\frac{d^2\omega}{dP^2}$) obtained from best quadratic fit to the measured data, the results of bulk modulus (B_0), isothermal volume compressibility (χ) and Grüneisen parameter (γ) for (60+5)- and (60+20)-SiNWs. The data for bulk Si are taken from Weinstein et al. [81] and Mernagh et al. [85].

Sample	ω_0	$\frac{d\omega}{dP}$	$(\frac{d^2\omega}{dP^2}) \times 10^{-2}$	B_0	χ	γ
	cm^{-1}	cm^{-1}/GPa	(GPa cm)$^{-2}$	GPa	GPa^{-1}	
(60+5)-SiNWs	518.7(4)	6.4(3)	−18(4)	80(5)	0.01253(9)	0.98(7)
(60+20)-SiNWs	520.5(2)	5.1(1)	−6(2)	100(5)	0.01002(6)	0.98(5)
Bulk Si[a]	519.5(8)	5.2(3)	−7(2)	98	0.0102	0.98(6)
Bulk Si[b]	518.6	5.5	−0.86	94.8	0.01055	1.0058

[a] Weinstein et al. [81]. [b] Mernagh et al. [85].

compared to its Bulk counterpart ($B' = 4.16$ [6]). However, the same comparision results for (60+20)-SiNWs show a 10% lower B' (3.72). Therefore, there is a sharper increase of the bulk modulus B of (60+5)-SiNWs with the applied pressure, causing a larger negative quadratic pressure coefficient (see Figure 5.12).

5.3.0.2 Raman linewidth of (60+5)- and (60+20)-SiNWs under hydrostatic pressure

In this section, the pressure dependence of the phonon linewidth of (60+5)- and (60+20)-SiNWs are discussed. The FWHM of the Lorentzian component was extracted by fitting a Voigt profile to the experimental data using a fixed width (2.5 cm^{-1}) for the Gaussian instrumental component (obtained from a line spectrum from a neon lamp, see Figure 5.14(b)). In Figure 5.14, the FWHM of (60+5)-SiNWs and bulk Si is plotted against the applied hydrostatic pressure. The filled and open circles refer to Raman linewidth as a function of increasing and decreasing pressure at room temperature, respectively. The results were reversible. The red solid line represents the calculated results of the bulk Si pressure dependence at $T = 0\,\text{K}$, taken from Menendez et al. [90]. The FWHM of SiNWs (filled and open circles) are larger than those of the bulk Si by a factor of 2 or higher. As discussed before

5.3 The effect of oxide layers on the vibrational properties of silicon nanowires

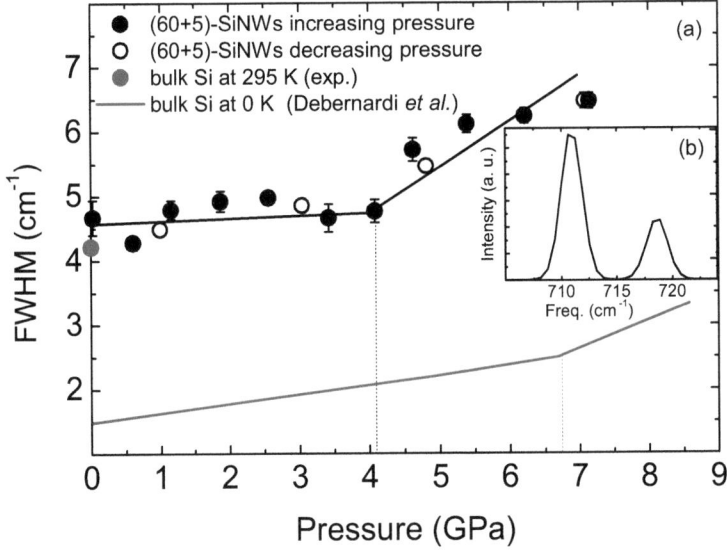

Figure 5.14: (a) Pressure dependence of FWHM of the Raman peaks of (60+5)-SiNWs and bulk Si. The filled (open) symbols denote data obtained under increasing (decreasing) pressure. The red solid line corresponds to the calculated results of the bulk Si pressure dependence of Raman line at $T = 0$ K, taken from Debernardi et al. [89]. The red solid circle refers to the measured FWHM of bulk Si at room temperature with the same Raman setup. (b) Part of the neon spectrum measured with the 514.5 nm excitation laser line. The FWHM of these peaks was measured to be about 2.5 cm^{-1} and used for the deconvolution of the spectra.

Chapter 5. Pressure dependence of the vibrational properties of SiNWs

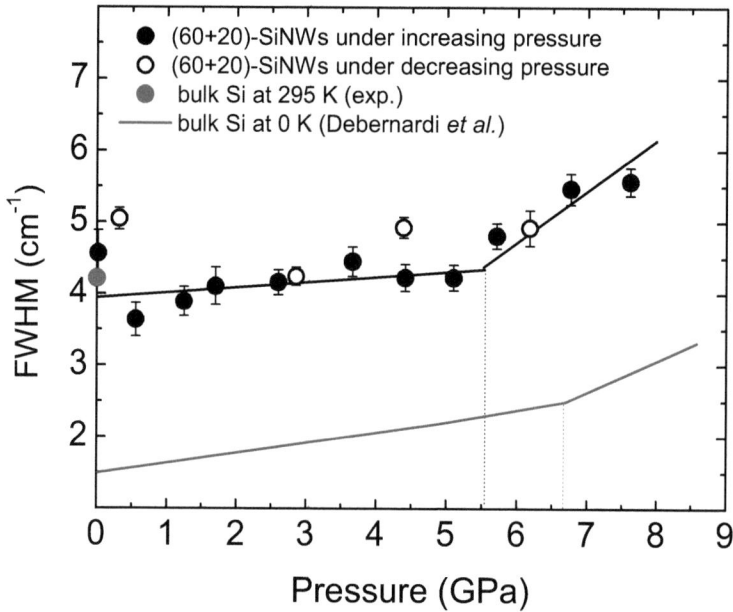

Figure 5.15: Pressure dependence of FWHM of the Raman peaks of (60+20)-SiNWs and bulk Si. The filled (open) symbols denote data obtained under increasing (decreasing) pressure. The red solid line represents the calculated results of bulk Si pressure dependence of Raman line at $T = 0\,\text{K}$ taken from Debernardi et al. [89]. The red solid circle refers to the measured FWHM of bulk Si at room temperature using the Raman setup, discussed in Section 5.2.1.

5.3 The effect of oxide layers on the vibrational properties of silicon nanowires

(Section 5.2.2.4), this difference is due to the fact that the calculation was made for $T = 0\,\text{K}$. Since the FWHM difference between SiNWs and the bulk Si at 0 K (red solid line in Figure 5.14) is about $2.5\,\text{cm}^{-1}$, the observed FWHM shift is merely a temperature effect and is not due to confinement. This notion is confirmed by the fact that for SiNWs with an average diameter of about 70 nm, the confinement effects are negligible, as reported by Piscanec *et al.* [29] and discussed in Chapter 4. The FWHM of bulk Si measured at room temperature using the Raman setup presented in Section 5.2.1 is about $4\,\text{cm}^{-1}$ (red solid circle in Figure 5.14), which further supports that the observed FWHM shift shown in Figure 5.14 results from the temperature effect. The increase of FWHM of the Raman peak with the increasing pressure originates from the pressure-induced changes in the phonon dispersion relation, which is related to the decay process of LTO into LO+TA phonons, as discussed before (Section 5.2.2.4). Beyond a critical pressure, an additional decay channel becomes active (LTO→LO+TA) causing a steep incline in FWHM.

From the measured line widths of (60+5)-SiNWs, shown in Figure 5.14, one can infer that the decay process of the LTO-phonon into the LO+TA -phonon becomes allowed at about 4 GPa, where a change in the slope of the FWHM pressure dependence is observed. In Figure 5.15, the FWHM of (60+20)-SiNWs is plotted as a function of the applied pressure. Here, the rise of FWHM occures at somewhat higher pressure of about 5.5 GPa. The activation of the additional decay channel at different pressures, for bulk Si (7 GPa), (60+5)-SiNWs (4 GPa) and (60+20)-SiNWs (5.5 GPa), suggests that the pressure affects the compressibility of SiNWs and the thickness of the SiO_x shell, and regulates the decay into LO + TA phonons. In other words, the pressure changes the dispersion relation of the bulk Si and SiNWs. The dispersion relations under pressure in the case of (60+5)- and (60+20)-SiNWs are not identical. Due to these differences, the final state of the decay, defined by the momentum and the energy conservation rules, in the case of (60+5)- and (60+20)-SiNWs differs from that of the bulk Si. Consequently, the decay channel into LO + TA phonons activates in different external pressures for (60+5)-, (60+20)-SiNWs and their bulk counterpart.

Chapter 5. Pressure dependence of the vibrational properties of SiNWs

5.3.0.3 Conclusions

In conclusion, two SiNW samples with different oxide layer thicknesses were investigated under applied pressures of up to 8 GPa under hydrostatic conditions, using Raman spectroscopy in a diamond anvil cell. The results revealed that the isothermal compressibility, which is the inverse of the bulk modulus, is approximately 20% higher and 2% lower for SiNWs with a thin and thick oxide shells, respectively, compared to the bulk Si. Moreover, the pressure derivative B' of (60+5)-SiNWs (with a thin oxide shell) is about 50% higher and for (60+20)-SiNWs (with a thick oxide shell) is about 10% lower than its bulk counterpart. The remarkable increase in FWHM of the Raman peak, which is interpreted as the consequence of the LTO-phonon decay into LO+TA phonons, starts at a lower pressure (4 GPa) for the SiNWs than in the case of the bulk Si (at about 7 GPa). These findings suggest that the thickness of the oxide layer of SiNWs affects the pressure coefficient of SiNWs, which regulates the value of the critical pressure for the activation of a new decay channel.

5.4 Strain in Si-SiO$_x$ core-shell nanowires

The LTO Raman frequencies of (60+5)- and (60+20)-SiNWs showed red and blue frequency shifts, respectively, compared to their bulk counterparts (see Table 5.4). These frequency shifts were originated from the lattice expansion and contraction of Si cores of (60+5)- and (60+20)-SiNWs, respectively, which in turn were caused by the tensile and compressive stresses, induced by their SiO$_x$ shells. It seems that the thickness of the oxide layer regulates the the type of stress (tensile versus compressive) in the SiNWs. To simulate the dependence of stress on the thickness of oxide layer in the SiNWs, calculations based on the linear continuum elasticity model are presented in the following section.

During the oxidation process of SiNWs, as the oxide grows, on the one hand the Si-SiO$_x$ interface advances into the silicon core. Consequently, the Si-SiO$_x$ interface is pulled inwards. On the other hand, SiO$_x$ increases its volume during that process,

5.4 Strain in Si-SiO$_x$ core-shell nanowires

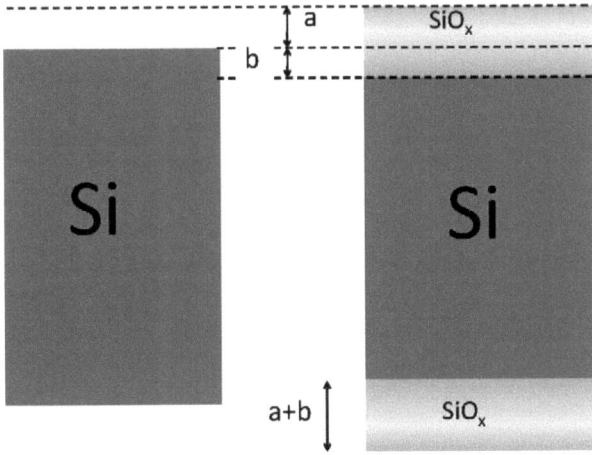

Figure 5.16: Consumption of Silicon during thermal oxidation. a and b are about 54% and 44% of the total oxide thickness, respectively.

which pushes the outer surface of SiO$_x$ outwards. Thus the generation of SiO$_x$ takes place in both directions relative to the original surface, as shown in Figure 5.16. In this figure a and b share about 54% and 44% of the total oxide thickness, respectively. Due to the stoichiometric relationships and the difference between the densities of Si and SiO$_x$, about 46% of the silicon surface is consumed during oxidation [103].

5.4.1 Strain in linear continuum elasticity model

In this model the crystal is considered an elastic body [104]. Each point of this body is identified by a vector $\vec{r}$ (unstrained). Applying a stress onto the body leads this vector to shift to a new position $\vec{r}'$. The displacement vector $u(\vec{r})$ is defined as $u(\vec{r}) = \vec{r}' - \vec{r}$. In order to describe the stress-strain relationship, the linear continuum mechanical model (Equation 5.14 in Section 5.1.2) was used. The total strain energy in this model is given by:

Chapter 5. Pressure dependence of the vibrational properties of SiNWs

$$U_{\text{cm}} = \frac{1}{2} \sum_{i,j,k,l} C_{ijkl} \epsilon_{ij} \epsilon_{kl}, \tag{5.36}$$

where U is minimized for a given three-dimensional structure, using finite differences for the strains $\epsilon_{ij} = \partial u_i / \partial x_j$ with u being the displacement vector field. The compliances C_{ijkl} are represented by the parameters C_{11}, C_{12} and C_{44} for cubic crystals, as discussed in Section 5.1.2 (Equation 5.14). It must be noted that the nonlinear effects as well as plastic deformations resulting from exceedingly large strain energies are not considered here.

For a body in equilibrium, the divergence of the stress tensor must be zero:

$$\nabla \cdot \sigma_{ij} = 0. \tag{5.37}$$

This equation is used in the linear continuum elasticity model. Moreover, in core-shell structures, the lattice distortion of the interface is not originated from externally applied stress but from lattice mismatches between the core and the shell. This conditions must be incorporated into the model. As an example, an interface between two materials with lattice constants $a^{(c)} = a_x^{(c)} = a_y^{(c)} =$ and $a^{(s)} = a_x^{(s)} = a_y^{(s)}$ in xy-plane can be considered. In this case the the conditions on stress of the interface is given by:

$$\left(1 + \epsilon_{xx}^{(c)}\right) a^{(c)} = \left(1 + \epsilon_{xx}^{(s)}\right) a^{(s)} \quad \text{and} \quad \left(1 + \epsilon_{yy}^{(c)}\right) a^{(c)} = \left(1 + \epsilon_{yy}^{(s)}\right) a^{(s)}. \tag{5.38}$$

These conditions ensure that the strain lattice constant of the shell matches that of the core at the core-shell interface. In order to implement these condition in the model, another displacement vector must be defined ($\tilde{u}(\vec{r})$). This vector is defined by a lattice constant a_M and the stress tensor $\tilde{\epsilon}_{ik}$ can be written as:

$$\tilde{\epsilon}_{ik} = \frac{a_M}{a}(\delta_{ik} + \tilde{\epsilon}_{ik}) - \delta_{ik}. \tag{5.39}$$

Typically, the coordinate system in which the stress, the strain (Equations 5.13) and the total strain energy (Equations 5.36) are described are $x \parallel [100]$, $y \parallel [010]$, $z \parallel [001]$. Using this coordinate system for SiNWs oriented in [111] direction, leads to some disadvantages such as waste of computational resources, problems with

5.4 Strain in Si-SiO$_x$ core-shell nanowires

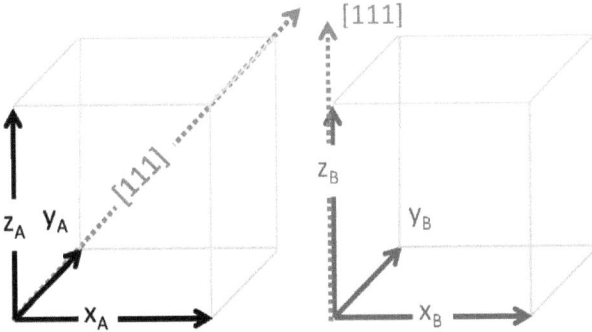

Figure 5.17: Coordinate system A corresponds to $x_A \parallel [100]$, $y_A \parallel [010]$, $z_A \parallel [001]$. [111] (green dashed vector) is the growth direction of SiNWs. The coordinate system B is defined as $x_B \parallel \left[\frac{1}{\sqrt{6}}, \frac{1}{\sqrt{6}}, -\sqrt{\frac{2}{3}}\right]$, $y_B \parallel \left[-\frac{1}{\sqrt{2}}, \frac{1}{\sqrt{2}}, 0\right]$, $z_B \parallel \left[\frac{1}{\sqrt{3}}, \frac{1}{\sqrt{3}}, \frac{1}{\sqrt{3}}\right]$. In this coordinate system the [111]-direction, which is the growth direction of SiNWs is parallel to the z axis.

boundary conditions for strain calculations, and difficulties to implement structures beyond the rotational symmetry. To overcome the disadvantages and to pave the way for studying SiNWs oriented in [111] direction, the stiffness tensor needs to be transformed, as outlined in the following section.

5.4.2 Transformation of stiffness tensor

Figure 5.17 shows the normal coordinate system (A) and the transformed coordinate system (B). As shown in this figure, the SiNWs (growth direction [111]) in coordinate system A were oriented in $[x_A, y_A, z_A]$ direction. The new coordinate system was defined such that the SiNWs were oriented in $[0, 0, z_B]$ direction. Consequently, the coordinate system A was transformed to the new (B) coordinate system B, according to:

Chapter 5. Pressure dependence of the vibrational properties of SiNWs

$$x_B \parallel \left[\frac{1}{\sqrt{6}}, \frac{1}{\sqrt{6}}, -\sqrt{\frac{2}{3}}\right], \quad y_B \parallel \left[-\frac{1}{\sqrt{2}}, \frac{1}{\sqrt{2}}, 0\right], \quad z_B \parallel \left[\frac{1}{\sqrt{3}}, \frac{1}{\sqrt{3}}, \frac{1}{\sqrt{3}}\right]. \quad (5.40)$$

The stiffness tensor, which in coordinate A is defined by Equation 5.13, is calculated for the transformed coordinate B using a rotation matrix:

$$R = (x_B, y_B, z_B)^T \quad (5.41)$$

The transformed stiffness tensor is given by:

$$C^B_{ijkl} = \sum_{mnop} R_{im} R_{jn} R_{ko} R_{lp} C^A_{mnop}. \quad (5.42)$$

This tensor in the B-basis C^B_{ij} yields:

$$C^B_{ij} = \begin{pmatrix} C^B_{11} & C^B_{12} & C^B_{13} & 0 & C^B_{15} & 0 \\ C^B_{12} & C^B_{11} & C^B_{13} & 0 & -C^B_{15} & 0 \\ C^B_{13} & C^B_{13} & C^B_{33} & 0 & 0 & 0 \\ 0 & 0 & 0 & C^B_{44} & 0 & -C^B_{15} \\ C^B_{15} & -C^B_{15} & 0 & 0 & C^B_{44} & 0 \\ 0 & 0 & 0 & -C^B_{15} & 0 & C^B_{66} \end{pmatrix}, \quad (5.43)$$

5.4 Strain in Si-SiO$_x$ core-shell nanowires

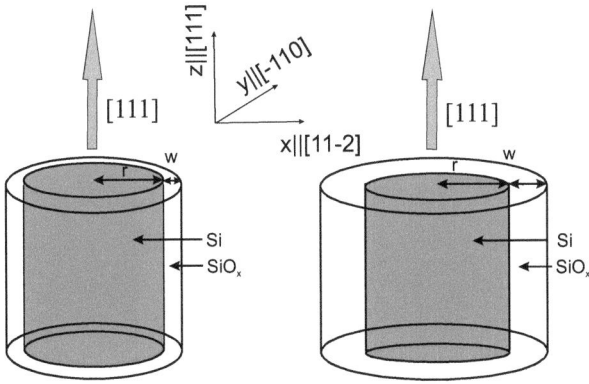

Figure 5.18: The schematic diagram of SiNWs, for which the stress/strain were calculated by means of the linear continuum elasticity model. The SiNWs were grown in [111]-direction. In these calculations w corresponds to the SiO$_x$ shell thickness. It takes the values of 5 nm and 20 nm for (60+5)-SiNWs and (60+20)-SiNWs, respectively. The core diameter r for both structures is 60 nm.

where the matrix elements in the B coordinate system C_{ij}^B are given by:

$$\begin{aligned}
C_{11}^B &= \frac{1}{2}(C_{11}^A + C_{12}^A + 2C_{44}^A) \ , \\
C_{12}^B &= \frac{1}{6}(C_{11}^A + 5C_{12}^A - 2C_{44}^A) \ , \\
C_{13}^B &= \frac{1}{3}(C_{11}^A + 2C_{12}^A - 2C_{44}^A) \ , \\
C_{33}^B &= \frac{1}{3}(C_{11}^A + 2C_{12}^A + 4C_{44}^A) \ , \\
C_{15}^B &= -\frac{1}{3\sqrt{2}}(C_{11}^A - C_{12}^A - 2C_{44}^A) \ , \\
C_{44}^B &= \frac{1}{3}(C_{11}^A - C_{12}^A + 4C_{44}^A) \ , \\
C_{66}^B &= \frac{1}{6}(C_{11}^A - C_{12}^A + 4C_{44}^A) = \frac{1}{2}C_{44}^B \ .
\end{aligned} \qquad (5.44)$$

Chapter 5. Pressure dependence of the vibrational properties of SiNWs

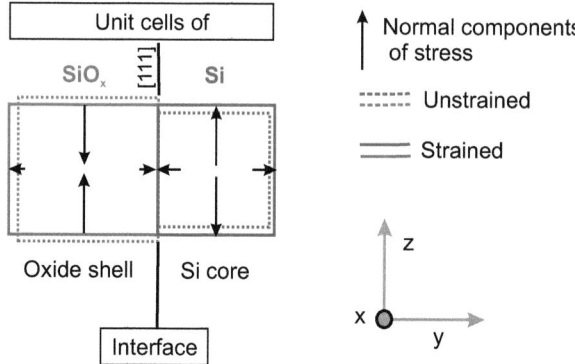

Figure 5.19: The schematic diagram of the interface between Si-core and SiO$_x$-shell in yz plane. The arrows represent the normal components of the stress. The unstrained and strained cases are shown with dashed and solid lines, respectively.

Having transformed the stiffness tensor, the stress σ_{ij} (Equation 5.13) and the energy of the stress U_{cm} (Equation 5.36) can be calculated in the B-basis.

5.4.3 Strain and the strain energy in Si-SiO$_x$ core-shell nanowires

The SiNWs studied in this work consist of a silicon core and a SiO$_x$ shell, as shown in Figure 5.16. The parameters used for the Si-core are: $C_{11} = 166$ GPa, $C_{12} = 64$ GPa, $C_{44} = 79.6$ GPa. In the case of the SiO$_x$-shell, the elastic parameters to be used are quite uncertain. Since the SiO$_x$-shell of the SiNWs increases in volume during the oxidation process. This can be modeled using a larger virtual lattice constant than that of the silicon core ($a_0^{\text{SiO}_x} = c \cdot a_0^{\text{Si}}$, $c = 1.1$). In doing so, the growing oxide shell exerts a tensile strain on the remaining Si-core, both in growth direction as well as in lateral directions perpendicular to the growth plane.

The structures investigated with this model were the (60+5)-and (60+20)-SiNWs, described in Chapter 2.3.2. A schematic diagram of these SiNWs are shown in Fig-

5.4 Strain in Si-SiO$_x$ core-shell nanowires

ure 5.18. The SiNWs were grown in [111]-direction. w and r correspond to the SiO$_x$ shell thickness and the core diameter, respectively. The different lattice constants are accounted for by using appropriate interface conditions as shown in Figure 5.19, where the lattice constant of the shell is assumed to be 10% larger than that of the Si-core. Figure 5.20 and 5.21 show the calculated strain (ϵ_{xx} and ϵ_{zz}) of (60+5)-SiNWs (green curve) and (60+20)-SiNWs (blue curve) as a function of radial position. The oxidation starts with a thin oxide layer exerting a low level of tensile stress on the remaining Si-core. Most of the strain energy density is accumulated in the shell with values surpassing 1.3 GPa (see Figure 5.22). As shown in Table 5.5, the stress levels are highest in growth direction (σ_{zz}) and tangential to the interface (σ_{xx}), and can be relaxed by a material redistribution process that reduces the strain components ϵ_{zz} and ϵ_{xx}. This likely happens during the complete process of oxidation. The only component that remains untouched is $\epsilon_{yy}/\sigma_{yy}$, which describes the radial strain/stress component. This redistribution process, however, effectively increases the radial projection of the shell's lattice constant leading to an increase of the exerted radial tensile stress on the Si-core. The thicker the shell, the more prominent this effect is, though it is not captured by the linear strain model applied here. The direction of the plastic relaxion is driven by a minimization of the strain energy inside the oxide shell: $\sigma_{xx}\epsilon_{xx} + \sigma_{yy}\epsilon_{yy} + \sigma_{zz}\epsilon_{zz} + \textit{off diagonal terms}$. For the thin shell, the radial contribution $\sigma_{yy}\epsilon_{yy}$ is an order of magnitude smaller than that for the SiNWs with the thick shell.

Figure 5.23(a) shows the Si-SiO$_x$ interface in SiNWs prior to oxidation, where two neighboring Si-unit cells exist, having the same lattice constant. With the start of oxidation process and the incorporation of oxygen, the size of the unit cells, shown in Figure 5.23(b), increases significantly, causing a lattice mismatch in the interface of Si-SiO$_x$. The resulting lattice mismatch can either lead to elastic or plastic relaxation. Figure 5.23 depicts the possible scenarios, which are explained in the following:

- Elastic relaxation (Figure 5.23(c)) leaves the SiO$_x$-Si interface coherent, meaning that Si atoms, which were neighbors prior to oxidation, remain neighbors across the interface even after oxidation. Such situations in the linear regime

Chapter 5. Pressure dependence of the vibrational properties of SiNWs

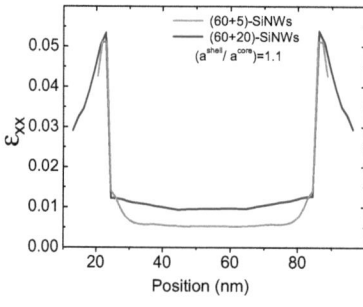

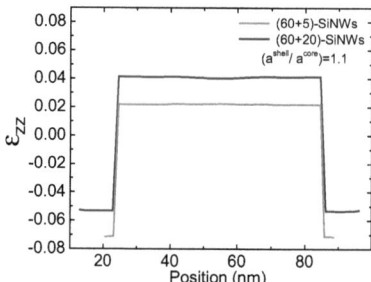

Figure 5.20: The strain (ϵ_{xx}) of (60+5)-SiNWs (green curve) and (60+20)-SiNWs (blue curve) as a function of position.

Figure 5.21: The strain (ϵ_{zz}) of (60+5)-SiNWs (green curve) and (60+20)-SiNWs (blue curve) as a function of position.

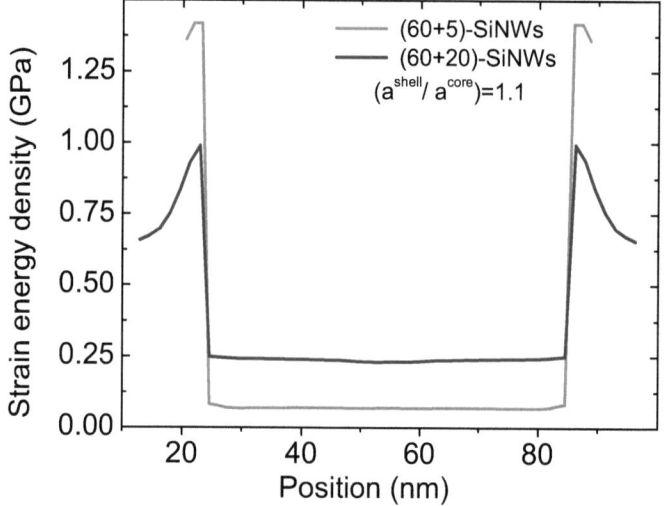

Figure 5.22: The strain energy density of (60+5)-SiNWs (green curve) and (60+20)-SiNWs (blue curve) as a function of position.

Table 5.5: Results for stress, strain and energy density at the boundary shell/core for (60+5)-SiNWs and (60+20)-SiNWs.

	(60+5)-SiNWs			(60+20)-SiNWs		
	SiO_x-shell	Si-core	Δ	SiO_x-shell	Si-core	Δ
σ_{xx}	−1.77	0.148	1.918	−1.36	0.47	1.83
σ_{yy}	0.2	0.4	0.2	0.4	0.53	0.13
σ_{zz}	−1.56	0.56	2.12	−1.09	1.04	2.13
ϵ_{xx}	−0.089	−0.002	0.087	−0.073	0.01	0.74
ϵ_{yy}	0.051	0.016	0.035	0.052	0.013	0.039
ϵ_{zz}	−0.068	0.025	0.093	−0.049	0.046	0.056
$\sigma_{xx}\epsilon_{xx}$	0.15753	−0,000296	−	0.09928	0.0047	−
$\sigma_{yy}\epsilon_{yy}$	0.0102	0.0064	−	0.0208	0.00689	−
$\sigma_{zz}\epsilon_{zz}$	0.10608	0.014	−	0.05341	0.04784	−

were captured by the linear elasticity model employed in this work. From this model's viewpoint, the shell exerts a tensile stress on the Si-core, both in growth direction [111], as well as in radial direction. No matter how thick the shell becomes, the resulting strain inside the core is tensile in both directions, which contradicts the experimental observations in case of the thin shell.

- Plastic relaxation (Figure 5.23(d)), whether it is creep, viscoelastic relaxation or fracture, changes the positions of neighboring Si-atoms irreversibly. Thus, parts of the stress are relaxed and the interface becomes incoherent compared to the original Si-atom positions.

  Here, the strain inside the Si-core results from a subtle interplay between the plastically relaxed stress along [111] and the mounting tensile stress in radial direction due to the material redistribution during the oxidation process.

On the basis of elastic and plastic relaxation processes discussed above, the effect of thin- and thick-oxide shell on the stress/strain can be explained as followed:

- Thin SiO_x shell:
  Plastic relaxation occurs in directions of the largest stress, which is at the

Chapter 5. Pressure dependence of the vibrational properties of SiNWs

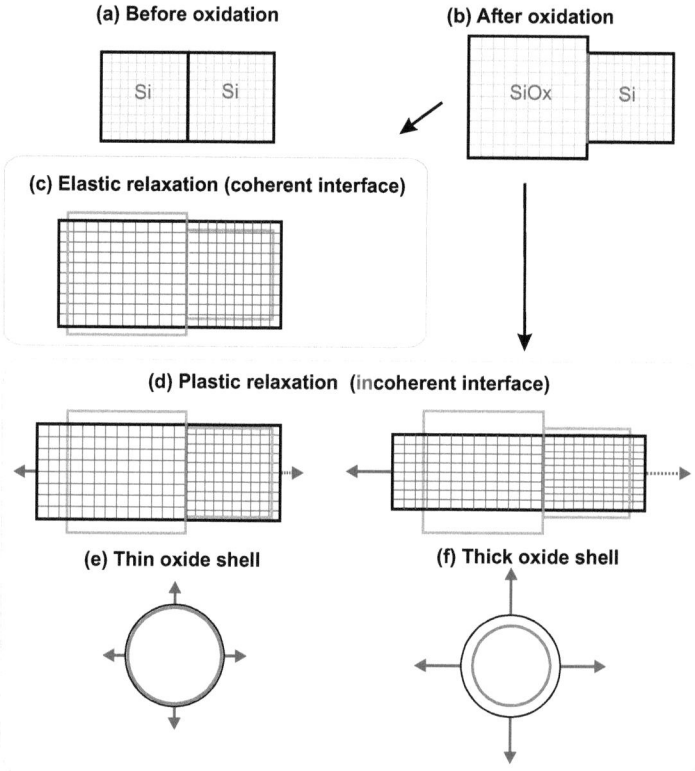

Figure 5.23: Schematic illustration of the Si-SiO$_x$ interface of SiNWs.

oxidation front in vertical (z or [111]) direction, due to the substantial lattice mismatch resulting from the oxygen incorporation. It can be suggested that the oxidized material at the interface redistributes in a way that the vertical stress component σ_{zz} becomes relaxed to a point where the strain energy density is low enough to reach the elastic regime.

If the vertical stress σ_{zz} was to be relaxed completely (as in the case of fracture), only the radial tensile stress onto the core would remain. In this hypothetical case, the radial strain components inside the core would be tensile, whereas the vertical components would be compressive due to the Poisson effect. This, however, contradicts the experimental observation for the thin shell, where the vertical strain is found to be tensile. Hence, σ_{zz} is not reduced completely and still exerts a limited tensile stress onto the core in the growth direction. The tensile stress results in the lattice expansion of SiNWs, leading to the observed higher compressibility of (60+5)-SiNWs compared to the bulk Si (see Section 5.3.0.1).

- Thick SiO_x shell:
  The thicker the shell the larger the radial tensile stress exerted onto the core becomes. The vertical stress, on the other hand, cannot exceed a certain value until it relaxes plastically. The radial stress component becomes larger with every step of oxidation process, due to incorporation of oxygen and the material redistribution towards the surface. The contribution of the latter to the radial stress is not captured by the elastic model.
  From the core's viewpoint, the radial tensile stresses from the shell increase as a function of shell thickness, whereas the vertical tensile stress remains constant. Now the Poisson effect comes into play: if a unit volume of a crystal is forced to become larger in one direction it responds by becoming thinner in the orthogonal directions unless other external force prevent it. Here, the radial tensile stresses become larger favoring a radial extension and a vertical compression of the Si-core unless vertical stresses exist. The vertical tensile stresses, however, are not scaling with shell thickness, but stay constant. At a certain shell thickness they are not capable anymore to compensate the Poisson effect and the core finnaly becomes compressively strained along the vertical

Chapter 5. Pressure dependence of the vibrational properties of SiNWs

direction, as observed experimentally in Section 5.3.0.1.

5.4.4 Conclusion

In oxidation process of SiNWs, the volume increase of the oxide shell was modeled using a virtual lattice constant larger than that of the Si core. The high pressure Raman measurements were accompanied by calculation of the stress-strain relation of the core-shell system in a linear continuum elasticity framework. In the case of the thin SiO_x shell, the strain energy is almost completely stored within the shell region, rendering the applied model valid for thin oxide shells. The stress-strain calculations show that in the case of the SiNWs with a thin oxide shell, the stress in growth direction is tensile. This explains the higher value of the lattice constant and consequently the higher compressibility of SiNWs with a thin oxide layer ((60+5)-SiNWs). As the shell becomes thicker, the strain energy stored inside the core seems to increase up to a point, where even nonlinear elasticity models become invalid and plastic deformation of the shell region is assumed to occur. The very strong counteracting stress components at the core-shell interface led to the conclusion, that during oxidation, the strain energy in this region is relaxed by incoherent atomistic reordering of the SiO_x molecules. This relaxation results in a compressive stress, which causes the lower compressibility of SiNWs with a thick oxide shell, compared to its bulk counterpart ((60+20)-SiNWs).

There are two possible outcomes: if the result confirms the hypothesis, then you've made a measurement. If the result is contrary to the hypothesis, then you've made a discovery.

Enrico Fermi

6
Vibrational properties of SiNWs: temperature dependence

An important part of applied research is studying the temperature dependence of the physical properties of materials. In terms of future applications, it is of particular interest to determine the expected properties of new materials in various temperatures. Temperature induces isotropic or anisotropic deformations in crystals. Heating the materials causes, on the one hand, an increased population of different phonon levels for each mode, and on the other hand, the expansion of the lattice parameter. Moreover, the decay of a certain phonon into two or three phonons is affected by the temperature, which causes an increase of the FWHM of the Raman mode.

Here, a detailed study of the temperature dependence of the Raman spectrum of SiNWs is presented. The temperature dependence of the first-order Raman peak

Chapter 6. Vibrational properties of SiNWs: temperature dependence

of SiNWs[29–31, 105, 106] and bulk Si [38, 90, 92, 107, 108] have been previously reported. In multiphonon Raman scattering, optical phonons close to the Γ-point as well as other phonons with larger wavevectors can participate (since the combination of two phonons with opposite wavevectors can always satisfy the fundamental Raman selection rule). Thus, additional information on the vibrational properties of SiNWs can be obtained from second-order scattering.

6.1 Theoretical approach: temperature-dependent Raman scattering

As the phonons are Bosons, they obey the Bose-Einstein distribution, which defines the behavior of phonons at various temperatures. This distribution gives the mean number of phonons n for any frequency ω at a given absolute temperature T :

$$n(\omega, T) = ((\exp \hbar\omega/k_B T) - 1)^{-1} \qquad (6.1)$$

where k_B and $\hbar$ are the Boltzmann's and Planck's constants, respectively. The number of phonons defines the intensity of the Raman scattering:

$$\frac{I_{AS}}{I_S} = \frac{I(\omega_i + \omega_0)}{I(\omega_i - \omega_0)} = \exp(-\hbar\omega/k_B T). \qquad (6.2)$$

To understand the effect of temperature on Raman frequencies and bandwidths of silicon, in particular SiNWs, some models have been developed [109, 110]. One of these models has been proposed by Balkanski et al. [92]. In an earlier work, based on numerical calculations of the Hardy-Karo deformation dipole model by Ipatova et al. [111], the temperature dependence of Raman frequencies and bandwidths were attributed to the anharmonic terms in the vibrational potential energy. When calculating the temperature dependence of the Raman modes, one can only consider the cubic anharmonicities in second-order. This leads to a damping constant (identifiable as FWHM of the Raman modes) proportional to the absolute temperature. Including the second-order quartic anharmonicity and/or the fourth-order cubic anharmonicity generates additional terms to the damping constant, which are proportional to T^2. As reported by Balkanski et al., the temperature dependence of the

6.1 Theoretical approach: temperature-dependent Raman scattering

Raman frequency and the damping constant of the LTO-phonons is applicable to the experimental data of the Raman spectra, in the range of 5 K to 1400 K, if cubic and quartic anharmonicities are included in the vibrational potential energies.

In order to study the temperature dependence of SiNWs and its deviation from its bulk counterpart, in the following section some models are introduced, which were used to interpret the temperature dependence of Raman frequencies. Subsequently, the data obtained from temperature-dependent Raman measurements in the context of these models will be compared with the results of Bulk Si in order to be able to make a statement about the vibrational properties of SiNWs.

6.1.1 Balkanski model

As the vibrational potential includes anharmonic terms, the generated optical phonons can decay into low-energy phonons. When the temperature increases, the decay processes intensify, resulting in an increased FWHM of the Raman peak. Moreover, the Raman peaks shift to lower energies, as mentioned before. According to Balkanski et al. [92], the frequency of the first-order optical Raman mode as a function of temperature has the form:

$$\omega(T) = \omega_0 + \Delta(T) = \omega_0 + A\left[1 + \frac{2}{e^x - 1}\right] + B\left[1 + \frac{3}{e^y - 1} + \frac{3}{(e^y - 1)^2}\right], \quad (6.3)$$

where ω_0 is the phonon frequency at 0 K, $x = \hbar\omega_0/2k_BT$ and $y = \hbar\omega_0/3k_BT$. The second term of Equation 6.3 characterizes the two-phonon processes, while the third term describes both the three-[1] and four-[2] phonon processes. A and B are constants, representing the weighting of each phonon process. According to Balkanski et al., the FWHM ($\Gamma(T)$) of the optical first-order Raman Peak can be written as:

$$\Gamma(T) = C\left(1 + \frac{2}{e^x - 1}\right) + D\left[1 + \frac{3}{e^y - 1} + \frac{3}{(e^y - 1)^2}\right] \quad (6.4)$$

---
[1] second-order cubic anharmonicity
[2] second-order quartic anharmonicity

Chapter 6. Vibrational properties of SiNWs: temperature dependence

where C and D are constants. The first and second terms of this equation are proportional to T and T^2, respectively. This model was found to be consistent with the experimental data obtained from the temperature dependence of the first-order optical Raman mode of silicon.

6.1.2 Cui model

Another model describing the temperature dependence of the first-order optical Raman modes is attributed to Cui et al. [112] and is based on the electron–phonon interactions in crystals. Cui et al. proposed an empirical formula to describe the temperature dependence of the Raman line position:

$$\omega(T) = \omega(0) - \frac{E}{\exp\left(F\hbar\Omega_0/k_BT\right) - 1}, \qquad (6.5)$$

where ω_0 is the Raman frequency at 0 K and, E and F are empirical constants. The origin of this equation is the temperature-related renormalization of the bandgap, when considering the electron-phonon interaction reported by Vina et al. [113]. It was supposed that the temperature-related changes of the gap, originating from electron-phonon interaction, must manipulate the phonon energies through the same mechanism. Equation 6.5 assumes that in temperature-related Raman frequencies, the electron-phonon interaction is just as important as the phonon-phonon interaction, which was reported by Balkanski et al. [92].

6.1.3 Mishra model

Mishra et al. [114] suggested a model to explain the temperature dependence of the first-order Raman frequency of silicon nanocrystals. This model is a combination of the phonon confinement-related RCF model, which was discribed in Chapter 4.1, and the model suggested by Balkanski et al. [92] related to the anharmonic phonon processes, which was discussed in Section 6.1.1. The intensity of an optical Raman

6.2 Raman scattering measurements of SiNWs: temperature dependence

peak can be obtained by incorporating the temperature dependent Raman frequencies (Equation 6.3) and the FWHM (Equation 6.4) in Equation 4.8:

$$I(\omega) = \int \frac{q dq \exp\left(-q^2 L_1^2/16\pi^2\right)}{\left(\omega - ([A + B\cos(q\pi/2)]^{0.5} + D) + \Delta(T) + \Delta_1\right)^2 + \left((\Gamma + \Gamma_1)/2\right)^2}. \quad (6.6)$$

The Δ_1 and Γ_1 are the Raman frequency shift and linewidth, respectively, due to the phonon confinement and lattice stress effects. The Raman frequencies and linewidths of SiNWs can be written as:

$$\omega(q, T) = \omega(q) + \Delta(T) + \Delta_1 \quad (6.7)$$

$$\Gamma(T) = \Gamma + \Gamma_1 \quad (6.8)$$

6.2 Raman scattering measurements of SiNWs: temperature dependence

6.2.1 Experimental details

The (8+3)-SiNWs were grown using the high-yield vapor transport technique [49], as explained in Chapter 2.3.3. The SiNWs were sonicated in isopropanol and dispersed on a high-grade steel. The Raman measurements were performed using a compact single-grating Raman spectrometer (Dilor LABRAM) equipped with microscope optics (OLYMPUS MD PLAN 10×) and a Peltier cooled CCD detector. To avoid laser heating, the 632.8 nm line of a He-Ne laser was focused with the microscope objective, which had only ten times magnification and the excitation power was kept below 1 mW. The experimental setup is shown in Figure 6.1.

A heating stage cryostat (LINKAM THMS 600) allowed temperature dependent measurements. The heating stage worked in the temperature range of −196°C to 600°C, with a temperature stability of 0.1° C. A highly polished pure silver heating

Chapter 6. Vibrational properties of SiNWs: temperature dependence

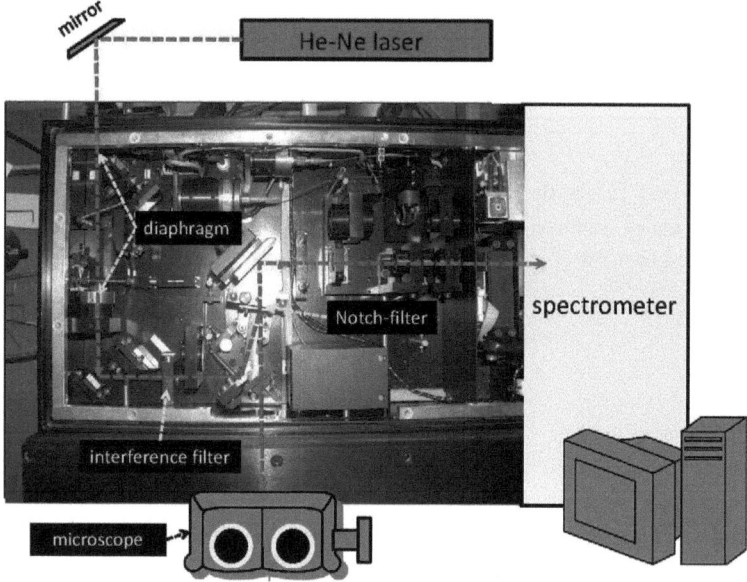

Figure 6.1: The experimental setup used for the temperature-dependent Raman measurement, consisting of a compact Raman spectrometer (Dilor LABRAM) equipped with microscope optics and a Peltier cooled CCD detector. The excitation wavelength is the 632.8 nm line of a He-Ne laser.

element ensured the heating, while the temperature was measured by a platinum resistor sensor. The cooling system consisted of a control unit housing, pumps and a two-liter dewar. The samples were cooled using liquid nitrogen. The controlled nitrogen-flow, in combination with the heating elements, determined the temperature of the sample. Warmed recycled dry nitrogen gas was used to purge the sample chamber and keep the window surface of the upper lid clear of condensation.

6.2 Raman scattering measurements of SiNWs: temperature dependence

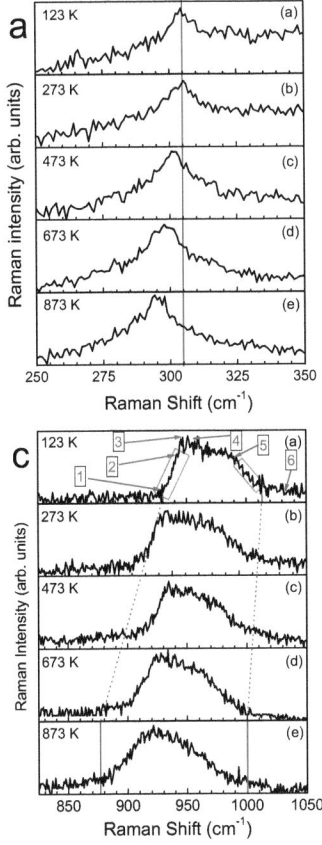

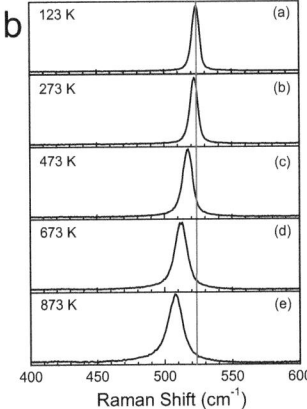

Figure 6.2: Raman Spectra of SiNWs as a function of temperature. (a) second-order transverse acoustic 2TA (X) peak. (b) first-order optical Raman mode at Γ. (c) second-order optical Raman band; 1-6 denote the 2TO at X, Q, $S1$, W, L and Γ points, respectively.

6.2.2 Results and discussion

Figures 6.2 plots the Raman spectra as a function of temperature, in the range of 123 K-873 K. Figure 6.2(b) shows that as the Raman peak shifts towards lower frequencies, its linewidth becomes larger. This effect has been reported previously [29, 105] and is explained as anharmonic phonon processes [29, 90, 92] combined with phonon confinement [92], as introduced in Section 6.1.1. In the case of the homogeneous heating with a heating stage, a symmetric broadening and a downshift of the Raman optical mode is observed, with no overall change of the line-shape. Figure 6.2(a) plots the evolution of the second-order acoustic band (2TA(X)) [115], which reflects the evaluation of the phonon density of states (DOS). The peak shifts from 303.7 cm^{-1} at 123 K to 292 cm^{-1} at 876 K, while the FWHM changes from 15 cm^{-1} to 32 cm^{-1}. Figure 6.2(c) shows the temperature dependence of the 2TO Raman band. This Raman band consists of features related to the two-phonon Raman scattering at various critical points, as shown in Figure 6.2(c). The numbers 1 to 6 denote the contribution from the X, Q, $S1$, W, L and Γ points, respectively [115, 116]. Figure 6.3 shows the 2TO Raman features of bulk Si. The high symmetry points are illustrated in BZ of Si (Figure 6.4). As the temperature increases, an overall downshift and broadening of this band occurs. Since this band is a mixture of 2TO peaks in different critical points and the linewidth of each peak is not distinguishable, we read out the zero-intensity crossing of the two shoulders (highlighted with green rectangles in Figure 6.2(c)) after a background subtraction.

Figure 6.5 plots the Raman frequency of the first-order optical mode as a function of temperature. For comparison, the Raman frequencies of the bulk Si are also plotted. These trends can be understood considering anharmonic phonon processes, as discussed in Section 6.1.1 (Equation 6.3). As shown in this figure, the Raman frequencies of (8+3)-SiNWs are lower than those of the bulk Si. The extracted fit parameters for (8+3)-SiNWs and the bulk Si are listed in Table 6.1. For (8+3)-SiNWs, frequency values $\omega_0 = 527.6$ cm^{-1} or $\omega(0) = 524.6(1)$ cm^{-1} exhibit an offset of about 1.4 cm^{-1} compared to those of the bulk Si ($\omega_0 = 529$ cm^{-1} or $\omega(0) = 525.8(1)$ cm^{-1}). This difference has two components: one due to heating from the laser (~ 0.4 cm^{-1}), and another due to phonon confinement (~ 1 cm^{-1} [29]), consistent with previous findings in Chapter 4.1. The constants A and B are the weighting factors related to the three- and four-phonon processes, which vary proportional to T and T^2, respectively. In the case of the three-phonon process, the weighting factor for (8+3)-SiNWs ($A = -3.02(1)$) is about 2% higher than that of the bulk Si ($A = -2.96(1)$). In four-phonon process, however, the opposite trend is observed, with the weighting factor for (8+3)-SiNWs ($B = -0.167$) being about 2% lower that its bulk counterpart

6.2 Raman scattering measurements of SiNWs: temperature dependence

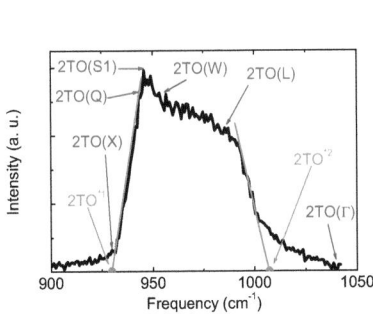
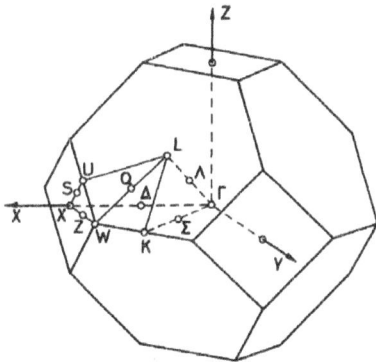

Figure 6.3: Raman spectrum of silicon in the region from $900\,\text{cm}^{-1}$ to $1050\,\text{cm}^{-1}$. The Raman features of X, Q, $S1$, W, L and Γ high symmetry points (Figure 6.4), were marked (red arrows). The data (green points) 2TO*[1] and 2TO*[1] are the zero-intensity crossing of the two shoulders of 2TO features.

Figure 6.4: First BZ of Si lattice and the high symmetry points and lines, taken from Neil et al. [117].

(-0.174). It must be noted that the four-phonon process part in the temperature-related shift of Raman frequencies begins to contribute at the temperatures higher than 700 K [92]. Figure 6.5 also shows that the temperature dependence of SiNWs from 272 K to 873 K, and the bulk Si from 272 K to 700 K can be approximated with a linear function(red lines). The linear fit parameters are listed in Table 6.2. From these it can be derived that the temperature-dependent decrease rate of LTO-frequencies for (8+3)-SiNWs ($2.36 \times 10^{-2}\,\text{cm}^{-1}$) is about 2.6% higher than that of the bulk Si ($2.29 \times 10^{-2}\,\text{cm}^{-1}$). The linear fit parameters are essential for comparing the temperature dependence of TO(Γ) peaks with those of 2TO and 2TA(X).

Chapter 6. Vibrational properties of SiNWs: temperature dependence

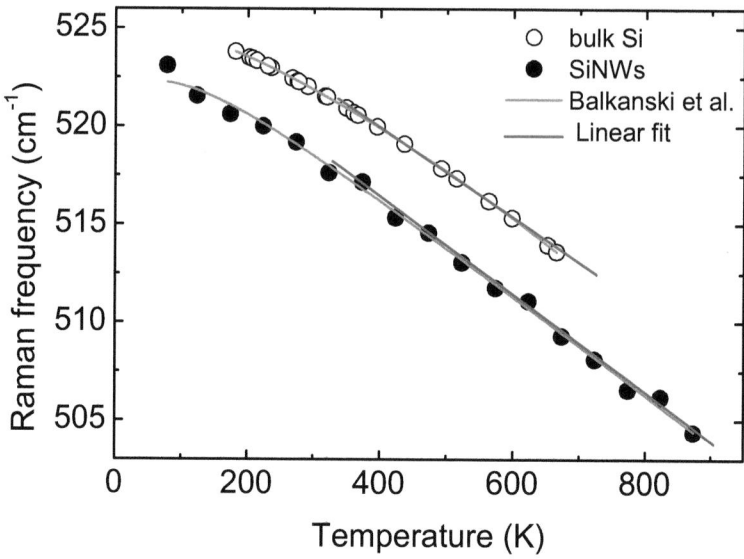

Figure 6.5: Raman frequencies of the first-order optical mode of (8+3)-SiNWs (filled circles) and bulk Si (open circles) as a function of temperature. The solid green curves are fits using three- and four-phonon processes [29, 92]. The red lines are linear fits to the data (see Table 6.2).

6.2.2.1 Interpretation of the results based on Balkanski *et al.* and Cui *et al.* models

Figure 6.6 shows the FWHM of (8+3)-SiNWs and bulk Si as a function of temperature. The dependence data is fitted with Equation 6.4 (Balkanski *et al.* [92]). The fit parameters are listed in Table 6.1. The Raman FWHMs of (8+3)-SiNWs show an offset of about $2 cm^{-1}$ compared to that of their bulk counterpart. This is due to the confinement-related red shift, as discussed in Chapter 4.1. However, the RCF model predicts an offset of $2.6 cm^{-1}$ for (8+3)-SiNWs compared to the bulk Si. The coefficients C and D, which are related to the three- and four-phonon processes, respectively, are also listed in Table 6.1. These confirm that the contribution of

6.2 Raman scattering measurements of SiNWs: temperature dependence

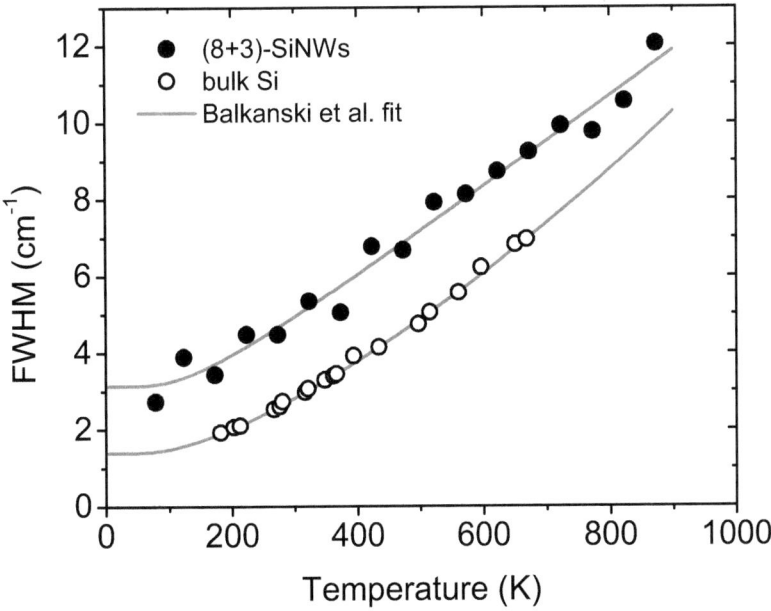

Figure 6.6: Linewidths of the first-order optical Raman mode of (8+3)-SiNWs (filled circles) and bulk Si (open circles) as a function of temperature. The solid green curves are fits using the Balkanski *et al.* model, based on three- and four-phonon processes [29, 92] (see Table 6.2).

the three-phonon processes ($C = 2.52(4)$) to anharmonic phonon processes is more prominent in the case of (8+3)-SiNWs compared to the bulk Si ($C = 1.29$), leading to a more pronounced temperature-dependent broadening.

Figure 6.7 shows the temperature dependence of the first-order Raman band of (60+5)-SiNWs and (60+20)-SiNWs. The fit data are extracted using Balkanski model and tabulated in Table 6.1 for comparison. The Raman frequency of (60+5)-SiNWs and (60+20)-SiNWs deviate from that of the bulk Si. It is noteworthy that the Raman scattering data for these samples were collected using the 488 nm line of an Ar$^+$ laser. The experimental macro-Raman set up (Chapter 5.2.1) was used.

Chapter 6. Vibrational properties of SiNWs: temperature dependence

Table 6.1: The extracted data from temperature dependent Raman measurement of SiNWs, on the basis of Balkanski et al. [92] and Cui et al. [112] models, and a linear fit.

Fit model	Fit parameter	unit	(8+3)-SiNWs	(60+5)-SiNWs	(60+20)-SiNWs	Si Bulk
Balkanski et al.	$\omega(0)$	cm^{-1}	524.6(1)	522.8(1)	524.4(1)	525.8
	A	cm^{-1}	−3.02(1)	−3.00(1)	−2.999(6)	−2.96
	B	cm^{-1}	−0.167(3)	−0.167(2)	−0.167(3)	−0.174
	Γ_0	cm^{-1}	2.52(5)	2.62(6)	2.30(5)	1.4
	C	cm^{-1}	2.52(4)	2.62(4)	2.30(5)	1.29
	D	cm^{-1}	0.00318	0.0004	0.0004	0.105
Cui et al.	$\omega(0)$	cm^{-1}	522.4(1)	520.6(1)	522.4(1)	524
	E	cm^{-1}	9.84(3)	9.90(6)	9.86(2)	9.51
	F		0.5046(4)	0.515(5)	0.513(1)	0.520
linear	$\omega(0)$	cm^{-1}	526.9	524.6(1)	525.29(6)	528.9
	$d\omega/dT \times 10^{-2}$	cm^{-1}/K	−2.36	−2.49	−2.28	−2.29

The excitation power on the sample was measured to be 2 mW. In this set up, the heating-related red shift in Raman frequencies due to the excitation power was determined to be 1.27 cm^{-1}/mW. It is concluded that excitation power-related red-shift of Raman frequencies of (60+5)-SiNWs and (60+20)-SiNWs is about 2.6 cm^{-1}/mW (dashed curves in Figure 6.7). From Table 6.1 it can be inferred that the anharmonic constants for three-phonon processes (A and C) are higher in the case of these SiNWs compared to those of the bulk Si, whereas the anharmonic constants of four-phonon processes (B and D) show the opposite trend. Figure 6.8 shows the linewidths of the first-order optical Raman mode of (60+5)-SiNWs (cyan solid circles) and (60+20)-SiNWs (blue solid circles). The higher FWHMs of (60+5)- and (60+20)-SiNWs compared to the bulk Si cannot be due to confinement, since the diameter of these SiNWs are about 60 nm. These observed inclines are due to the heating effect induced by laser excitation power, as well as the broadening of the

6.2 Raman scattering measurements of SiNWs: temperature dependence

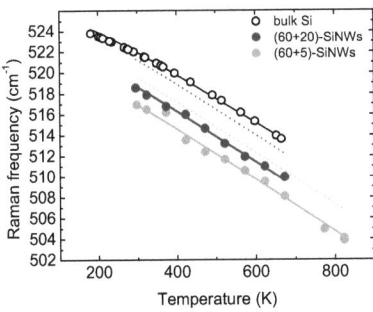
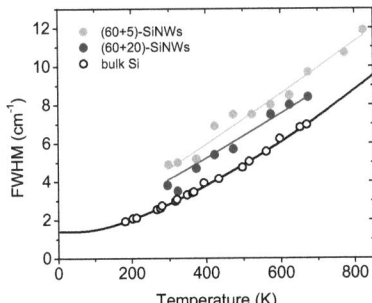

Figure 6.7: Raman frequencies of the first-order optical mode of (60+5)-SiNWs (cyan solid circles), (60+20)-SiNWs (blue solid circles), and the bulk Si (open circles) as a function of temperature.

Figure 6.8: Linewidths of the first-order optical Raman mode of (60+5)-SiNWs (cyan solid circles), (60+20)-SiNWs (blue solid circles), and bulk the Si (open circles) as a function of temperature.

Raman peak originated from the lattice expansion and contraction in (60+5)- and (60+20)-SiNWs, respectively, as discussed in Chapter 5.

As discussed in Section 6.1.2, Cui *et al.* [118] describe the temperature dependence of the first-order optical Raman modes on the basis of the electron-phonon interactions in crystals (Equation 6.5). The optical first-order Raman frequencies of (8+3)-SiNWs, (60+5)-SiNWs, (60+20)-SiNWs and the bulk Si are fitted using Equation 6.5. The extracted fit parameters (w_0, E, F) are shown in Table 6.1. Aside from the observed offset of the Raman frequency of SiNWs compared to the bulk Si, a more pronounced E for SiNWs is observed. The E coefficient is related to the linear part of the frequency decrement. The F coefficient, which determines the curvature of the function, is lower in the case of SiNWs compared to the bulk Si. It should be stressed that from E and F constants no insight into the physical properties of SiNWs can be gained.

The temperature dependence data for first-order Raman mode of SiNWs and the bulk Si indicate that anharmonic constants related to the peak width and position are higher in SiNWs, implying a higher degree of anharmonicity. The greater anhar-

Chapter 6. Vibrational properties of SiNWs: temperature dependence

monicity can be a confinement-related effect, which is the case of (8+3)-SiNWs. The lattice expansion induced by the SiO_x shell and resulting stress can be the origin of the increased anharmonicity. This is the case for (60+5)-SiNWs, which can not have confinement-related effects.

6.2.2.2 Temperature dependence of the 2TA(X) Raman band

The temperature-induced shift of the phonon frequencies is the sum of a frequency shift determined by the thermal volume expansion and the thermal population of the vibrational levels. One of the interesting points in lattice dynamics of Si is the negative frequency derivative with respect to the pressure of the TA phonons [81]. Consequently, the temperature-related softening of the 2TA(X) frequencies cannot be a result of the thermal volume expansion of Si, but instead originate from the change in the population of the vibrational levels. This leads to a change in average position of the atoms, resulting in anharmonocity of the potential [119].

For second-order Raman scattering studies as a function of temperature, the (8+3)-SiNWs were chosen. Figure 6.9 plots the Raman frequencies of the second-order acoustic features for (8+3)-SiNWs (filled circles) and bulk Si (open circles) as a function of temperature. The solid red lines represent linear fits to the experimental data. Table 6.2 lists the fit parameters. A red shift of the position of this peak with a rate of 1.57×10^{-2} cm^{-1}/K and a broadening of 2.19×10^{-2} cm^{-1}/K are observed for SiNWs. The 2TA(X) Raman frequencies of SiNWs are ~ 3 cm^{-1} lower than those of the bulk Si. This shows that the 2TA(X) band is also affected by the confinement. This is in contrast to the findings of Mishra et al. [120], who reported that the second-order acoustic band 2TA(X) is not influenced by the confinement effect and is similar to that of its bulk counterpart. As shown in Table 6.2, for SiNWs, the rate of temperature-induced frequency softening of the TO(Γ) mode (2.36×10^{-2} cm^{-1}/K) is about 50% higher than that of the 2TA(X) band (1.57×10^{-2} cm^{-1}/K). This is likely to be due to the fact that thermally-induced volume expansion does not cause a frequency red shift for the 2TA(X) band; but instead only the thermally-induced population change of the phonon levels are responsible for the softening of the 2TA(X) band. Comparing the rate of the softening of 2TA(X) of SiNWs

6.2 Raman scattering measurements of SiNWs: temperature dependence

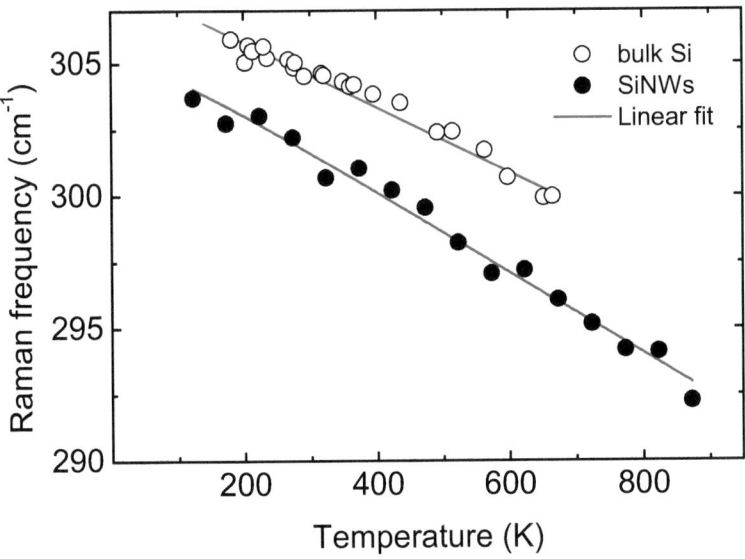

Figure 6.9: Raman frequencies of second-order peak 2TA(X) for SiNWs (black solid circles) and bulk Si (open circles) as a function of temperature. Red lines are fits to the data

with that of the bulk Si reveals that this rate is about 32% higher in the case of SiNWs. The anharmonic effect, which leads to a temperature-induced softening of the TA(X) band is more pronounced in SiNWs compared to its bulk counterpart.

6.2.2.3 Temperature dependence of the 2TO Raman band

As mentioned before, with increasing temperature a red shift and a broadening of the 2TO Raman band is observed. The line-widths of the individual peaks for each contributing phonon band cannot be determined. Figure 6.2(c) shows the 2TO band, composed of phonons belonging to the X, Q, $S1$, W, L and Γ critical points [116]. These critical points, in the case of SiNWs as well as bulk Si, were found to shift towards the lower energies as the temperature increased. The frequency softening for

Chapter 6. Vibrational properties of SiNWs: temperature dependence

SiNWs and bulk Si exhibits linearity. The frequencies of the 2TO at critical points were fitted with a linear function. The extracted data, namely the frequency of the critical points in 0 K, at ambient temperature, and the softening rate of the frequencies $(-d\omega/dT)$ for SiNWs and bulk Si are tabulated in Table 6.2. All of the critical point frequencies (except the frequency of 2TO(Γ)) of (8+3)-SiNWs were found to be lower than the corresponding values for the bulk Si. The critical points at 923 cm^{-1}, 942 cm^{-1}, 945 cm^{-1}, 948 cm^{-1}, 983cm $^{-1}$, and 1040 cm^{-1} in the bulk Si [116], appear at 919cm^{-1}, 933cm^{-1}, 939 cm^{-1}, 946cm $^{-1}$, 979 cm^{-1}, and 1041 cm^{-1}, respectively, in the case of (8+3)-SiNWs. It results that the 2TO band of (8+3)-SiNWs is influenced by the confinement effect and shifts towards lower frequencies. This frequency shift is not the same for the frequencies of all critical points, $i.e.$ the shifts for 2TO(X) is $\Delta\omega = 4$ cm^{-1}, for 2TO(Q) $\Delta\omega = 9$ cm^{-1}, for 2TO($S1$) $\Delta\omega = 6$ cm^{-1}, for 2TO(W) $\Delta\omega = 2$ cm^{-1} and for 2TO(L) $\Delta\omega = 2$ cm^{-1}. These results are in agreement with the confinement-related enhanced anharmonic effects in Si nanocrystals reported by Mishra $et\ al.$ [114, 120]. Table 6.2 also shows that the rate of the temperature-induced softening of the frequencies of the 2TO critical points is higher in the case of (8+3)-SiNWs compared to the bulk Si. One exception is the 2TO(Γ)(index 6 in Table 6.2 and Figure 6.2(c)), which shows a weaker dependence in SiNWs than in the bulk Si. At 300K the frequency of 2TO(Γ) for SiNWs is 1040cm^{-1}, whereas for bulk Si it is 1041cm^{-1}.

In order to study the temperature-related broadening of the 2TO band of (8+3)-SiNWs, the zero-intensity crossing of the two shoulders shown in Figures 6.2(c) and 6.3 were used. These are denoted as 2TO^{*1} and 2TO^{*2}. As shown in Table 6.2, the frequency of 2TO^{*1} (2TO^{*2}) for (8+3)-SiNWs at ambient temperature is determined as 916.6 cm^{-1} (1009 cm^{-1}), which is 7.7 cm^{-1} lower (higher) than this frequency for the bulk Si. The temperature-induced shift of the 2TO^{*2}-point towards higher frequencies results from the fact that with increasing temperature the 2TO band becomes broader. The broadening of the band counters the temperature-induced shift of the 2TO^{*2}-point towards the lower frequency, whereas in the case of the 2TO^{*1}-point the broadening amplifies the shift of the frequency to the lower values. In order to calculate the pure frequency shift of the 2TO^{*1}- and the 2TO^{*2}-

6.2 Raman scattering measurements of SiNWs: temperature dependence

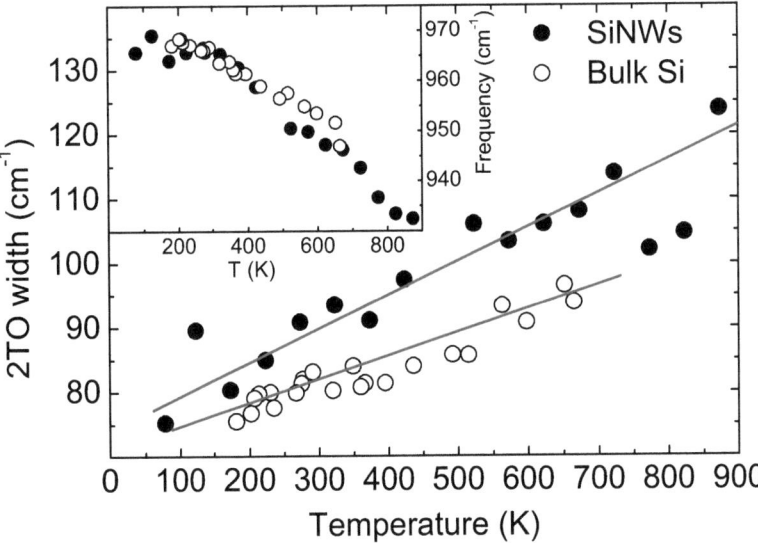

Figure 6.10: The difference of zero-intensity crossing values of the two shoulders of the 2TO density of states feature (width). Inset: The average of the zero-intensity crossing values of the two shoulders of the 2TO Raman feature.

points, the broadening part of the frequency shift should be removed. This is conditional upon the data concerning the width of the 2TO band, which will be discussed below.

Figure 6.10 shows the overall width of 2TO band as a function of the measured temperature. The widths were defined as the frequency difference ($\Delta\omega$) between 2TO^{*1} and 2TO^{*2}. The width of 2TO band increases linearly with temperature. As shown in Figure 6.10, the width of this band in ambient temperature is about 10 cm^{-1} higher for (8+3)-SiNWs, relative to the bulk Si, suggesting that the 2TO band-width of (8+3)-SiNWs is also influenced by confinement effects. Moreover, the rate of the temperature-induced broadening for (8+3)-SiNWs is found to be $(4.4 \pm 0.6) \times 10^{-2}$ cm^{-1}/K, which is about 26% higher than that of the bulk Si $((3.5 \pm 0.4) \times 10^{-2}$ cm^{-1}/K). These results are also consistent with the confinement-

Chapter 6. Vibrational properties of SiNWs: temperature dependence

related to the broadening of the 2TO band of nanocrystalline Si reported by Mishra et al. [114], who found a broadening of $8\,\text{cm}^{-1}$ for nanocrystalline Si with dimension of 7 nm.

To filter out the broadening effect from the frequency shift of the $2TO^{*1}$ and $2TO^{*2}$ points, it can be assumed that the temperature-induced broadening of the 2TO band is symmetrical and can be obtained from the slope of the width-temperature diagram in Figure 6.10. This value is $(4.4 \pm 0.6) \times 10^{-2}\,\text{cm}^{-1}/\text{K}$ for (8+3)-SiNWs and $(3.5 \pm 0.4) \times 10^{-2}\,\text{cm}^{-1}/\text{K}$ for the bulk Si. To determine the pure softening-rate of $2TO^{*1}$-frequency, half of the broadening rate ($0.026\,\text{cm}^{-1}/\text{K}$ for (8+3)-SiNWs and $0.017\,\text{cm}^{-1}/\text{K}$ for bulk Si) should be subtracted from the measured rate of the softening (see Table 6.2). The resulted rates of $2TO^{*1}$-frequency softening are $0.052\,\text{cm}^{-1}/\text{K}$ for (8+3)-SiNWs and $0.041\,\text{cm}^{-1}/\text{K}$ for bulk Si. In the case of $2TO^{*2}$ points, instead of the subtraction, an addition of the values should be carried out. This results in a softening rate of $0.0604\,\text{cm}^{-1}/\text{K}$ for (8+3)-SiNWs and $0.0412\,\text{cm}^{-1}/\text{K}$ for the bulk Si. The results are listed as $2TO^{*1c}$ and $2TO^{*2c}$ in Table 6.2. The results show, that the softening rate of the frequencies for $2TO^{*1}$ and $2TO^{*2}$ points are higher in the case of SiNWs compared to their bulk counterparts. The temperature-induced broadening also affects the position of $2TO^{*1}$ and $2TO^{*2}$ points at 300 K, which are listed in Table 6.2. At 300 K, the the temperature induced broadening is equal to $10.5\,\text{cm}^{-1}$ and $13.3\,\text{cm}^{-1}$ for (8+3)-SiNWs and the bulk Si, respectively. Half of this value must be subtracted from the frequency of $2TO^{*1}$ at 300 K and added to the frequencies of $2TO^{*2}$, in order to obtain the position of the $2TO^{*1}$ and $2TO^{*2}$ points. The corrected frequencies are tabulated in Table 6.2 as $2TO^{*1c}$ and $2TO^{*2c}$. The inset of Figure 6.10 plots the average position of the 2TO shoulders, derived as the center of the 2TO structure. As the temperature increases, the central point of the TO band is found to shift to the lower frequencies. The rate of this shift is measured to be $-0.059(6)\,\text{cm}^{-1}/\text{K}$ for (8+3)-SiNWs, which is about 60% higher than that of its bulk counterpart ($-0.037\,\text{cm}^{-1}/\text{K}$).

A comparison between the temperature-induced shift of $1TO(\Gamma)$-mode and the 2TO band shows that the rate of the softening of 2TO band is about twice higher than that of the $1TO(\Gamma)$-mode for (8+3)-SiNWs as well as the bulk Si. This is resulted from the fact that in the case of 2TO-mode, two phonons are involved and the

6.2 Raman scattering measurements of SiNWs: temperature dependence

Table 6.2: Temperature dependence of Raman frequencies of SiNWs and bulk Si. *1 and *2 denote the frequencies at the points where the shoulders of the 2TO density of states feature cross zero. These are highlighted with green rectangles in Fig.6.2c. 1-6 are the 2TO critical points, Fig.6.2c.

Raman modes and features	SiNWs			Bulk silicon		
	ω_0 cm^{-1} $T=0$K	ω cm^{-1} $T=300$K	$-d\omega/dT$ cm^{-1}/K $\times 10^{-2}$	ω_0 cm^{-1} $T=0$K	ω (cm^{-1}) cm^{-1} $T=300$K	$-d\omega/dT$ cm^{-1}/K $\times 10^{-2}$
1TO(Γ)	526.9	518.4	2.36	528.9	521.1	2.29
2TA(X)	306.5	301.6	1.57	308.2	304.6	1.19
2TO(X)^1	936(2)	919	5.82	940	923	5.0
2TO(Q)^2	947	933	4.60	954	942	3.76
2TO($S1$)^3	952	939	4.33	958	945	4.22
2TO(W)^4	961	946	4.92	962	948	4.63
2TO(L)^5	998	979	6.21	1001	983	5.76
2TO(Γ)^6	1052	1041	3.70	1053	1040	4.41
2TO^{*1}	943.8	916.3	7.94	942.1	924	5.87
2TO^{*2}	1024	1009	3.93	1012.6	1005.3	2.37
2TO^{*1c}	–	922.6	5.29	–	929.3	4.11
2TO^{*2c}	–	1002.3	6.04	–	934.5	4.12

resulting shift is approximated as the addition of the participating phonon-shifts.

6.2.2.4 The temperature effect on the intensities of 2TA(X) and 2TO bands

Figure 6.11 plots the relative intensities of SiNWs I(2TA(X)) and I(2TO) (normalized to the I(1TO)) as a function of temperature. The relative intensity of I(2TO) is higher than I(2TA(X)) for (8+3)-SiNWs and the bulk Si. With increasing temperature, both I(2TA(X)) and I(2TO) increase relative to the first-order TO. The absolute Raman intensity of the first-order optical Raman mode of SiNWs and the

Chapter 6. Vibrational properties of SiNWs: temperature dependence

Table 6.3: Temperature dependence of Raman relative intensities of SiNWs and bulk Si.

	$dI/dT \times 10^{-4}$		$I(T=0)$	
	SiNWs	Bulk Si	SiNWs	Bulk Si
2TA(X)/1TO(Γ)	1.9	2.57	0.0054	0.016
2TO/1TO(Γ)	0.9	1.62	0.119	0.115
2TA(X)/2TO	7.2	10.1	0.17	0.5948

bulk Si are reduced when the temperature increases, while the absolute intensities of both 2TA and 2TO increase. Table 6.3 lists the fitted data from Figure 6.11. In general, the Raman intensity depends on the value of the absorption coefficient, which in the case of Si increases with temperature [121]. Since photon absorption is a process which competes with Raman scattering, reduction in Raman intensities is expected to occur as the temperature rises. Moreover, the first-order Raman intensity (Raman scattering cross section) is proportional to $(n(\omega, T) + 1)$, whereas for two phonons the proportionality is given by $(n(\omega, T) + 1)^2$, $n(\omega, T)$ being the Bose-Einstein distribution function $(n(\omega, T) = 1/\exp(\hbar\omega/K_B T) - 1)$. Consequently, both I(2TA(X)) and I(2TO) (normalized to I(1TO)) will increase with increasing temperature.

The Raman relative intensities I(2TA)/I(2TO) increase with temperature for both SiNWs and the bulk Si, as shown in Figure 6.12. This means that the scattering efficiency of the acoustic phonon peaks increases more with temperature compared to that of the optical phonons. To understand the physics of this phenomenon one possible approach is to use the Hamiltonian of the Electron-phonon interaction. The second-order term of the Hamiltonian for interaction between energy band electrons and acoustic phonon is given as [52, 122]:

$$H_{int}^2 = -i \sum_k \sum_{q,\lambda} \left(\frac{N\hbar}{2M\omega_{q\lambda}} \right)^{0.5} + V_{q+k_n} \left(e_{q\lambda}(q+k_n) \right) \times (a_{q\lambda} + a^+_{-q\lambda}) C^+_{k'} C^+_k, \quad (6.9)$$

where k and q are wavevectors of electron and phonons. $e_\lambda(k)$ is the polarization vector of the λ mode, which satisfies the orthogonality conditions. $a^+_{-q\lambda}$ and $a_{q\lambda}$ are

6.2 Raman scattering measurements of SiNWs: temperature dependence

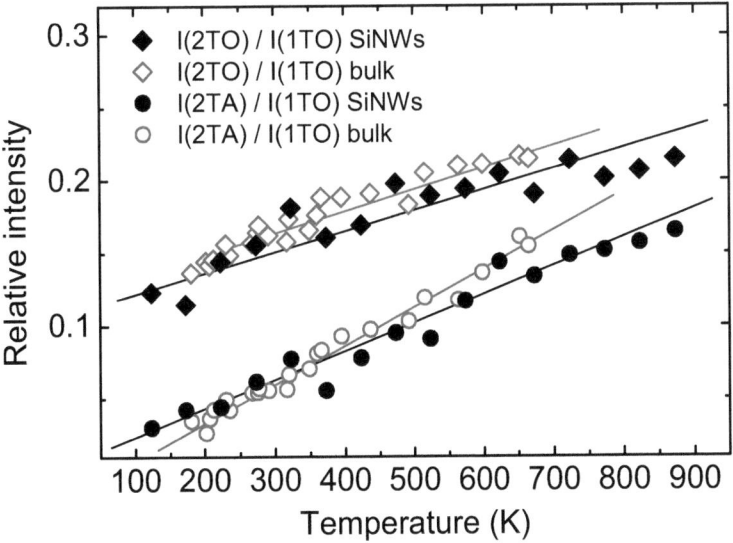

Figure 6.11: Relative (normalized to the 1TO) intensities of 2TA(X) (black filled circles) and 2TO (black filled diamonds) peaks of SiNWs. For comparison, the relative intensity of 2TA (gray open circles) and 2TO (gray open diamonds) for bulk Si are also plotted.

the creation and annihilation operators of a phonon with the polarization λ. V_{q+k_n} and $C_{k'}^+ C_k^+$ are the terms from the long range Coulomb interaction. The wavevector k' can be written as $k' = k + q + k_n$. Pursuant to the long wave approximation, $e_{q\lambda}$ can be split into a parallel and a perpendicular component in respect to the wavevector of phonon q. The parallel component $e_{q\lambda}^{LA}$ represents the LA and the perpendicular component $e_{q\lambda}^{TA}$ represents the TA phonons. For the case where $k+q$ is in the first BZ, $k_n = 0$ results in $e_{q\lambda}^{TA} = 0$ and consequently $e_{q\lambda}.(q+k_n) = e_{q\lambda}^{LA}$. This means that in this case, the TA phonons do not participate in the electron-phonon interaction process. In contrast, when $k+q$ exceeds the first BZ, $k_n \neq 0$ results in $e_{q\lambda}.(q+k_n) = e_{q\lambda}^{LA} + e_{q\lambda}^{TA}.k_n$. In other words, for the wavevectors out of the first BZ, the LA phonons as well as the TA phonons contribute to the electron-phonon

Chapter 6. Vibrational properties of SiNWs: temperature dependence

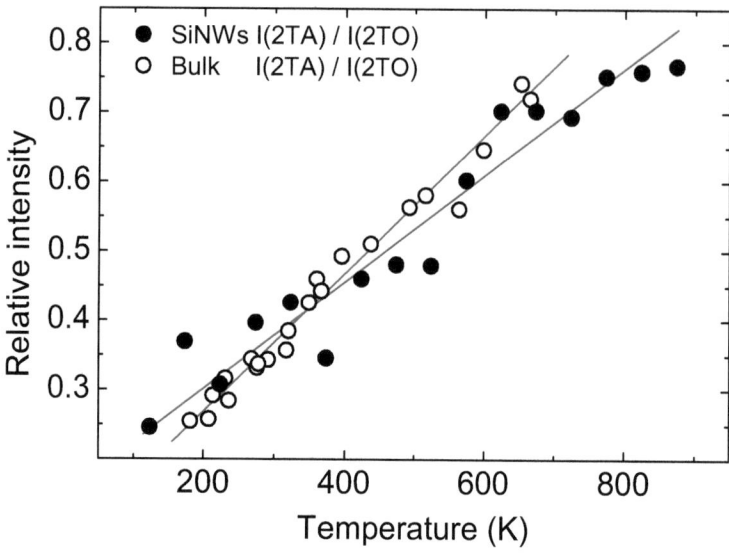

Figure 6.12: Intensities of 2TA phonons relative to 2TO phonons (I(2TA)/I(2TO)) for (8+3)-SiNWs (filled circles), as a function of temperature. For comparison, the corresponding data for bulk Si (open circles) is also plotted.

interaction process. With increasing temperature, the scattering of those phonons which have a high wavevector q and exceed the first BZ, become pronounced and consequently the contribution of $e_{q\lambda}^{TA}$ phonons increases. In contrast to the acoustic phonons, the optical phonons participate in the scattering process in both cases of $k_n = 0$ (q in first BZ) and $k_n \neq 0$ (q exceeds first BZ). From the discussion above, it can be inferred that with temperature rise the relative intensity of 2TA phonons to 2TO phonons should increase. The slopes of the relative intensity of 2TA to 2TO phonons (Figure 6.12) for (8+3)-SiNWs and the bulk Si are listed in Table 6.3. The rate of the relative intensity $d\left(I(2TA)/I(2TA)\right)/dT$ for SiNWs was found to be 7.2×10^{-4}, which is about 28% lower than its bulk counterpart (10.1×10^{-4}).

6.3 Conclusion

The temperature dependence of 1TO(Γ), 2TA(X) and 2TO Raman bands were investigated for the temperature range of 123 K to 873 K. The resulting red shift of the 1TO, 2TA(X) and 2TO peaks of (8+3)-SiNWs were found to be higher than those of the bulk Si. In SiNWs, the shift of 2TA(X) was found to be lower than that of the 1TO mode. For the case of 2TO, the temperature-induced shift was about twice higher than that for 1TO. Different dependencies of the scattering cross sections of first- and second-order peaks result in increase of I(2TO)/I(TO) and I(2TA)/I(TO) with increasing temperature. In particular, with increasing temperature the 2TA cross section increases relative to the 2TO. The temperature dependence of Raman features of SiNWs indicate a more prominent anhamonicity in SiNWs. This can be due to the size effect, which was the case in (8+3)-SiNWs, or due to the lattice expansion induced by the SiO_x shell and resulting stress, which was the case for (60+5)-SiNWs.

In physics, you don't have to go around making trouble for yourself, nature does it for you.

Frank Wilczek

7

Excitation power dependence on the vibrational properties of SiNWs

In this chapter a detailed study of the Raman spectrum of (15+5)-SiNWs and its dependence on the excitation laser power is presented. The results of the excitation power dependence of the Raman spectrum of (15+5)-SiNWs are compared with their Raman scattering data at various temperatures. The main difference between these two sets of experiments was that in the case of excitation laser power dependent measurements, the (15+5)-SiNWs were heated inhomogeneously (locally), so that their temperature had a Gaussian distribution profile. In contrast, studying the temperature dependent Raman spectra of (15+5)-SiNWs (Chapter 6) dealt with the homogeneously heated sample.

Chapter 7. Excitation power dependence on the vibrational properties of SiNWs

7.1 Experimental details

(15+5)-SiNWs were grown using the vapor transport technique, as described in Chapter 2.2 [49]. The macro-Raman system was the same setup as that described in Chapter 5.2.1. All data were obtained with the 514.5 nm line of an Ar^+ laser. The focused laser spot had a diameter of about 50 μm. The signal was detected in back-scattering geometry with a spectral resolution of 1 cm^{-1}. For the laser excitation power dependent measurements in vacuum and air an Oxford Instruments cryostat with a single sapphire window was used. The (15+5)-SiNWs sample was mounted on a cold finger. For the Raman measurements in vacuum, the volume surrounding SiNWs was evacuated to approximately 10^{-5} mbar, using a turbo molecular pump. A manometer was flanged to the sample chamber exhaust pipe. A heating-stage cryostat (LINKAM THMS 600), as described in Chapter 6.2.1, was used for temperature-dependent measurements. One of the important experimental aspects when conducting excitation power dependent measurements on any inhomogeneous sample, including SiNWs, is to guarantee the reproducibility of the laser spot on the sample. This was realized as follows: firstly, the aperture and laser current of Ar^+ laser were fixed during the experiment, in order to ensure a constant power density of the laser spot on the sample. Secondly, the power of the laser was adjusted using neutral density filters. The placement of these filters would lead to a change in the laser beam line, and consequently a modified focal point position as well as a shift of laser spot on the sample. In order to avoid this effect, a variable neutral glass filter wheel was used.

7.2 Results

In order to distinguish local heating induced by the laser from the widespread homogeneous heating, the shift and the intensity of the first optical Raman mode were monitored under two different conditions. The first set of Raman experiments were conducted in the ambient conditions and then in vacuum environment. Afterwards, Raman scattering experiments were conducted at various temperatures to investigate the differences between global and local heating on the Raman spectra of (15+5)-SiNWs.

7.2.1 Excitation power dependence measurements in ambient conditions

The Raman spectra of SiNWs were recorded at different laser excitation powers in the range of $1 - 50\,\text{mW}$, as shown in Figure 7.1. The spectra were fitted with a Lorentzian, plus a Gaussian for the low-energy tail [32, 75]. The resulting shifts and signal strengths are plotted in Figure 7.1. Starting at $\sim 517.5\,\text{cm}^{-1}$, the phonon frequency decreased initially, at a rate of $\sim -0.5\,\text{cm}^{-1}/\text{mW}$, with increasing power (black circles). Beyond the critical power $p_c \sim 10-15\,\text{mW}$ the slope almost vanished ($0.03\,\text{cm}^{-1}/\text{mW}$) and remains slightly positive, in agreement with earlier findings of Scheel et al. [32, 75]. In order to investigate the reversibility of this process, the power was then reduced gradually to zero (red circles). Under decreasing laser power, the phonon frequency returned to its original value at $517\,\text{cm}^{-1}$, though via a different path in the frequency-excitation power diagram. The hysteretic form is clearly apparent in this figure. The normalized Raman intensity, given by the area of the Raman signal divided by the excitation laser power, remained constant up to the critical power (p_c) when heating (Figure 7.1 (b)). Beyond p_c the intensity dropped drastically (by about a factor of 7) to $p = 45\,\text{mW}$. On the returning branch, the intensity remained at this low level and did not recover to its initial value (red circles in Figure 7.1 (b)).

This behavior can be understood by assuming a change in morphology of the sample. Upon laser heating, the (15+5)-SiNW bundle most likely contracted and moved out of focus, explaining the drastic intensity decrease beyond p_c. Correspondingly, the Raman shift ceased to change, remaining at $\sim 512\,\text{cm}^{-1}$. Upon lowering the power, the Raman peak returned to its original frequency, but the signal strength remained weak, due to the loss of focus. The large error bars in Figure 7.1(a) for the last four Raman frequencies under decreasing power show the fitting inaccuracy of the Raman frequency. Since the Raman frequency returned to its original

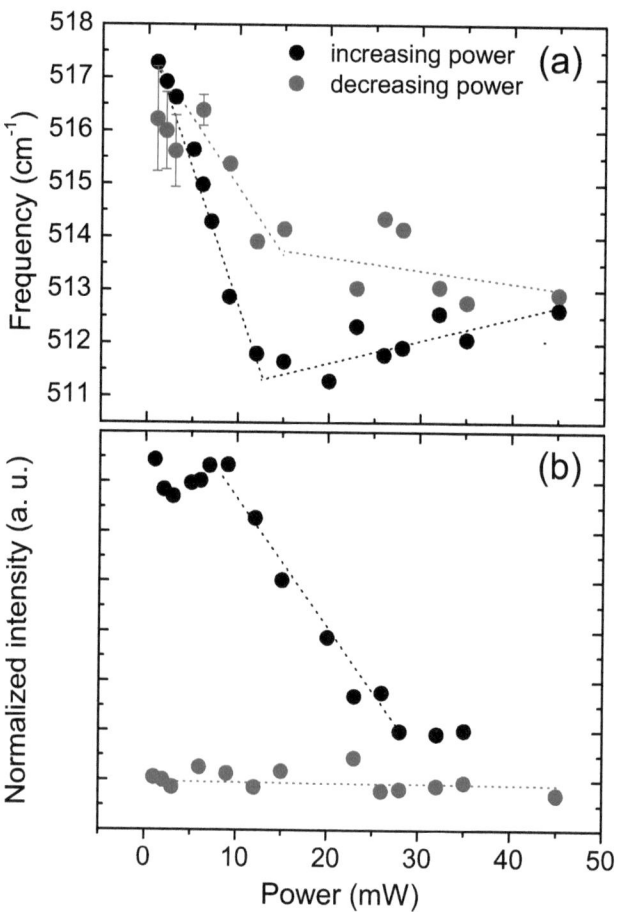

Figure 7.1: (a) (15+5)-SiNWs phonon frequencies as a function of the excitation power in air with increasing (black solid circles) and decreasing (red solid circles) power. (b) Normalized intensities as function of the excitation power in air. The normalized intensities were calculated as the Raman peak area divided by the laser power.

value, it can be concluded that the microscopic structure, *i.e.* (15+5)-SiNWs diameter, remained the same. Melting and re-solidification would lead to a more bulk-like frequency (520 cm^{-1} [115]) and therefore can be excluded. Moreover, as mentioned before, the laser spot diameter on the sample in the macro-Raman setup is about 50 µm. The power density in this case is about 100 times lower than that in a typical micro-Raman setup (spot diameter about 5 µm). A laser-induced local heating might also promote stress/strain in the SiNWs, due to a temperature gradient. Knowing that (15+5)-SiNWs are only a few micrometers long and the diameter of the laser spot on the sample is 50 µm, a strong temperature gradient on the SiNWs causing a destruction of wires, can be ruled out. Since the sample was not "refocused" under decreasing power, one can conclude that a macroscopic morphological change occurred, which might be explained as a vertical or sidewise contraction of the SiNWs network. To confirm this, power dependent measurements under increasing and subsequent decreasing excitation power were performed, where the laser was refocused after each measurement, so that the intensities during increasing and decreasing power were equal (see Figure 7.2). As shown in this figure, the Raman frequencies under increasing power are approximately the same as those under decreasing power. The slight difference between the frequency values under increasing and decreasing power most likely originated from the refocusing errors.

7.2.2 Excitation power dependence measurements (in vacuum)

Scheel *et al.* [32] have reported previously that the slope of the Raman shift versus the laser excitation power depends on the gas surrounding SiNWs [32]. Most of the heat from the SiNWs is removed via convective conduction by the surrounding gas, rather than by thermal contact between the SiNWs and the substrate [32]. In order to rule out differences in the gas thermal conductivity in the environment surrounding SiNWs *e.g.*, at the onset of the turbulent heat conduction, Raman excitation power dependent measurements were repeated in vacuum, at a pressure of 2×10^{-6} mbar. Figure 7.3 shows that the phonon frequency shift is larger in vacuum than in the air by a factor of 2, consistent with previous findings [32]. Here

Chapter 7. Excitation power dependence on the vibrational properties of SiNWs

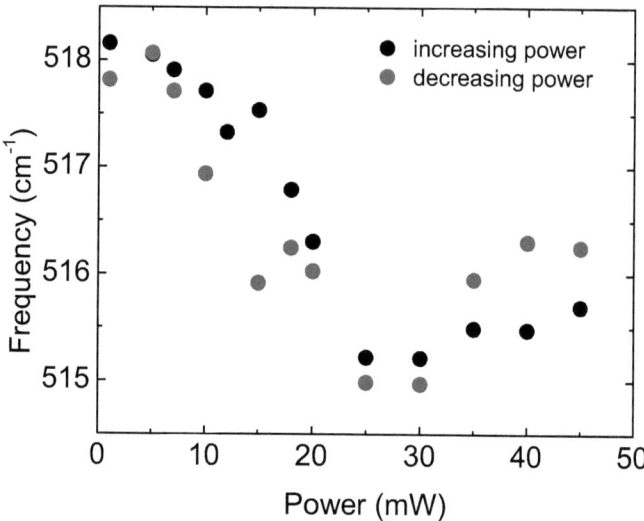

Figure 7.2: Phonon frequencies of (15+5)SiNWs as a function of the excitation power in air, under increasing power (black filled circles) and under decreasing power (red filled circles). The laser spot was refocused after recording each data point.

too, a hysteresis is observed upon power reduction, when returning to the original Raman shift of 518 cm^{-1} at zero power. It should be noted that the maximum difference in the hysteresis loop is ~ 2 cm^{-1} for the measurements in the air, and ~ 4 times larger in vacuum. From the occurrence of the hysteresis in vacuum, it can be concluded that the change in Raman slope is not due to a change in heat convection of the surrounding gas, rather it is caused by the sample moving out of focus during the inhomogeneous laser heating process. This interpretation is supported by the increase in hysteresis strength and confirmed by the normalized Raman intensity in vacuum, shown in Figure 7.3.

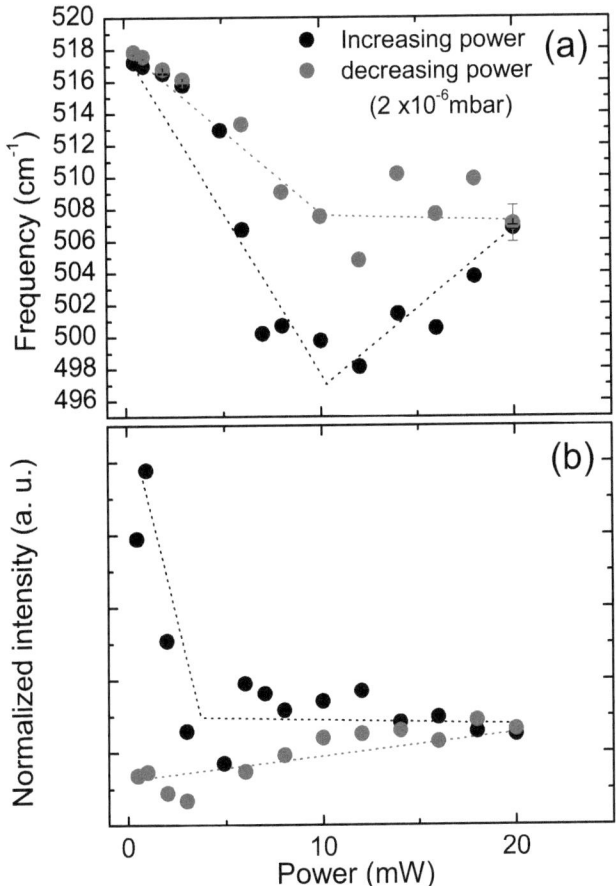

Figure 7.3: (a) (15+5)-SiNWs phonon frequencies as a function of excitation power in vacuum for increasing (black filled circles) and decreasing power (red filled circles). (b) Normalized intensities as a function of excitation power in vacuum. The normalized intensity was calculated as the Raman peak area divided by the laser excitaion power.

Chapter 7. Excitation power dependence on the vibrational properties of SiNWs

7.2.3 Homogeneous temperature dependent measurements

In order to further investigate the effects of heating, the (15+5)-SiNWs sample was placed on a heating stage and the Raman shift was recorded as a function of temperature. With this arrangement, the uniform heating of the (15+5)-SiNWs was realized, as opposed to the local laser induced heating. In order to examine the reversibility of the heating process, the Raman shift and intensity for rising and falling temperature were recorded. Figure 7.4 (a) plots the peak position as a function of the heating stage temperature, in the range of $300 - 900$ K, for increasing (black filled circles) and decreasing temperatures (red filled circles). The Raman frequency declines, with a uniform rate of ~ -0.02 cm/K, with increasing temperature (no change of the slope). The heating and cooling processes are reversible, as far as the peak position is concerned (no hysteresis). Any change in the microscopic structure would affect this reversibility. Figure 7.4 (b) shows the peak intensity as a function of temperature. The intensity drops for increasing temperature, and the values for decreasing temperature are much lower (by a factor of ~ 3) than those for increasing temperature. This can not be due to a microscopic structural change, because the intensity drops both, during the heating and cooling processes, but instead can be attributed to a macroscopic modification of the SiNWs network. Since the Raman frequency only depends on the homogeneous temperature, this change in morphology is reflected only in a decreasing Raman intensity. In the power-dependent Raman measurements the local temperature varies with a consequent change of morphology, resulting in the saturation and hysteresis observed in Figure 7.1(a).

7.3 Conclusion

In conclusion, a saturation of the Raman peak position in the power dependent measurements was observed. This saturation can not be attributed to a change from the laminar to turbulent gas flow in the environment, but rather to a morphological variation induced by the laser heating. This was confirmed by comparison of temperature-dependent and power-dependent (in vacuum) Raman measurements of (15+5)-SiNWs. It was shown that the saturation of the Raman frequencies, mea-

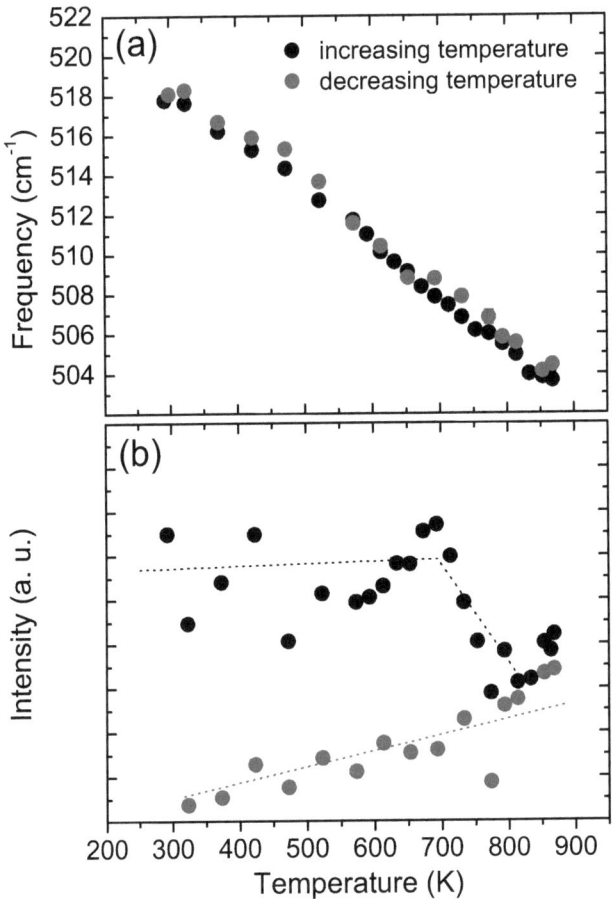

Figure 7.4: (a) SiNW phonon frequencies as a function of temperature, in air, measured with a heating stage for increasing (black circles) and decreasing temperature (red filled circles).(b) Raman intensity as a function of increasing (black filled circles) and decreasing (red circles) temperature in air.

Chapter 7. Excitation power dependence on the vibrational properties of SiNWs

sured for a bundle of SiNWs under increasing excitation power, is a consequence of the altered macroscopic morphology and is not due to a change in the convection or in the microstructure of the sample. Thus, by using high excitation power densities, macroscopic morphology changes can be induced in SiNWs, while retaining the microscopic structure.

Results! Why, man, I have gotten a lot of results. I know several thousand things that won't work.

Thomas A. Edison

8
Summary and outlook

In the present work, Raman spectroscopy was used to study the effects of confinement, hydrostatic pressure and homogeneous as well as inhomogeneous temperatures on the vibrational properties of SiNWs. The study of the Raman frequencies of (8+3)-SiNWs under hydrostatic pressure showed a more pronounced blue shift of the Raman frequencies relative to bulk Si. From the pressure coefficient of the Raman frequencies, a lower bulk modulus and a higher compressibility were calculated for (8+3)-SiNWs, compared to bulk Si. Analyzing TEM images revealed that the lattice constant of (8+3)-SiNWs was larger than its bulk counterpart. The lattice expansion was found to be the origin of the higher compressibility of (8+3)-SiNWs. It was found that the higher compressibility of (8+3)-SiNWs also affected the decay process of the optical phonons. In this regard, it is of particular importance to study this sample by means of high-pressure synchrotron diffraction experiments, in order to determine the lattice constant as a function of the applied pressure. More-

Chapter 8. Summary and outlook

over, a series of high pressure Raman measurements on single SiNWs with different diameters could provide valuable information about the dependence of the elastic parameter on the diameter of the SiNWs. The remarkable increase of the FWHM of the Raman first-order optical peak, which was interpreted to be a result of the LTO-phonon decay into LO+TA phonons, started at a lower pressure for the SiNWs than in the case of the bulk Si. This suggested that the different compressibility of SiNWs regulates the value of the critical pressure for the activation of a new decay channel.

Since in SiNWs a Si core is typically covered by a SiO_x shell, they can be presented as a core-shell structure. The results of Raman measurements on SiNWs with various thicknesses of the oxide shells under hydrostatic pressure showed that the thickness of the oxide shell modified the lattice constant, and consequently, the elastic properties of SiNWs. The thin SiO_x shell of (60+5)-SiNWs led to a tensile strain and thus to a lattice expansion, whereas the increase of the oxide shell thickness resulted in a compressive stress and a lattice contraction of the Si-core. In the case of (60+5)-SiNWs the lattice expansion of the Si-core was observed in the diffraction pattern of the TEM image. These measurements were accompanied by a calculation of the stress-strain relationship of the core-shell system, in a linear continuum elasticity framework. For thin SiO_x-shells, the strain energy is almost completely stored inside the shell region, rendering the applied model valid for thin oxide shells. In the case of thicker shells, the strain energy stored inside the core increases up to a point, where even nonlinear elasticity models become invalid and a plastic deformation of the shell region must be assumed. The very strong contracting stress components at the core-shell interface led to assume that during oxidation, the strain energy occurring in this region is relaxed by incoherent atomistic reordering of the SiO_x molecules, resulting in the suspension of the lattice expansion in the core. A capable continuation of high pressure Raman studies on SiNWs would be to search for the structural phase transition for nanowires with various diameter and oxide-layer thickness. Knowing that the compressibility of (8+3)- and (60+5)-SiNWs is higher than that of bulk Si, for SiNWs a lower phase transition pressure can be expected. One possible continuing study will be to determine the pressures at which the phase transitions happen.

The temperature-dependence of the first and second-order Raman peaks and features of SiNWs were investigated. The findings showed a more pronounced temperature coefficient for first-order optical Raman frequency for SiNWs compared to the bulk Si. In order to obtain more information concerning the vibrational properties, the second-order Raman features of SiNWs were studied as a function of temperature. These peaks were red-shifted as the temperature increased and the rate of the shifts was found to be higher compared to the bulk Si. The results showed, that the scattering efficiency of 2TA(X) and 2TO phonons rise, as the temperature increases. The rate of efficiency increase was found to be more pronounced for the acoustic mode, in comparison with the optical one. The origin of this increase was determined to be the increase of the cross section of the electron-phonon interaction, in the case of the acoustic modes at high temperatures. Thus, it is reasonable to evaluate the temperature dependence of the other Raman features: TA(X) and LO(L).

To determine the effect of inhomogeneous heating on the vibrational properties, the Raman optical mode of SiNWs were studied as a function of the laser excitation power. The excitation power having a Gaussian intensity distribution, induces a temperature gradient on the SiNWs. It was found that the observed saturation of the Raman frequencies, measured on a bundle of SiNWs under the increasing excitation power, was a consequence of morphological modifications of the SiNWs, and not due to a change in the convection regime or alternation of the micro-structure of SiNWs.

Recently, Lange *et al.* [123] have confirmed the existence of the radial breathing mode (RBM) in CdSe nanorods by Raman spectroscopy. The frequency of the RBM was found to be strongly dependent on the diameter of the nanorods. This RBM frequency is more sensitive to diameter fluctuations than that of the LO mode and consequently it is more adequate for an estimation of the nanorod diameter. Moreover in the case of core-shell nanorods the RBM frequency was observed to be affected by the thickness of the shell (ZnSe). Theoretical calculations reported by Li [124] have shown the presence of RBMs in SiNWs. The frequency of these modes was calculated to increase with the decreasing of SiNW diameter. A reasonable continuation of this work would be to experimentally confirm the existence of the RBMs in SiNWs and study their dependence on diameter. This approach would be used for a more sensitive estimation of the diameter of SiNWs. It would be of

Chapter 8. Summary and outlook

higher relevance to study the changes in frequency of RBMs in SiNWs at different pressures and/or temperatures, and to search for possible structural phase transition (from cubic to hexagonal) in SiNWs with various diameters. As mentioned before in case of CdSe nanorods the RBM frequency affects by the thickness of the shell (ZnSe). This effect is observable even if the shell consist of one or two monolayers, providing the opportunity to estimate the shell thickness more precisely in core shell nanowires.

Another important field in SiNWs research, which can be considered for future studies, is photoluminescence (PL) measurements. The PL spectrum of SiNWs exhibits double emission bands, which can be connected with defect states, amorphous phase and impurity-activated optical transmission [125]. The thickness of oxide coating of SiNWs seems to affect the PL Spectrum as discussed in Chapter 5.3. The PL measurements on SiNWs with various oxide shell thicknesses would provide more detailed insight into the role of oxide shell in photoluminescence properties of SiNWs. Moreover the study of PL under high pressures and at high temperature can provide valuable information about the structural properties of oxide-coated SiNWs.

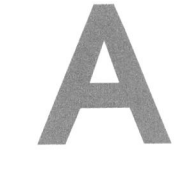

Appendix

A.1 Murnaghan equation of state

Bulk Modulus described the response of a system to an applied external hydrostatic pressure as:

$$B_0 = -V \left(\frac{\partial P}{\partial V}\right)_{V_0} \tag{A.1}$$

Pursuant the Murnaghan's equation of state [126], bulk modulus of materials varies linearly with the applied hydrostatic pressure:

$$B = c\left(1 + kP\right) \tag{A.2}$$

Chapter A. Appendix

where c and k are defined as:

$$c = -V\left(\frac{\partial P}{\partial V}\right)_0 \equiv B_0 \tag{A.3}$$

$$ck = -\frac{\partial}{\partial P}\left(V\frac{\partial P}{\partial V}\right)_0 \equiv B_0' \tag{A.4}$$

Integrating Equation A.1 yields:

$$P(V) = \frac{B_0}{B_0'}\left(\left(\frac{V_0}{V}\right)^{B_0'} - 1\right) \tag{A.5}$$

and

$$V(P) = V_0\left(1 + B_0'\frac{P}{B_0}\right)^{-1/B_0'} \tag{A.6}$$

or

$$a = a_0\left(1 + B_0'\frac{P}{B_0}\right)^{-1/3B_0'} \tag{A.7}$$

where a_0 and a are the lattice constants of crystals without and under applied external pressure, respectively.

A.2 Brich Murnaghan equation of state

Another way to describe the behavior of solids under pressure was presented by Francis Birch [127]. According to Maxwell's relations the pressure P is the derivative of F with respect to V:

$$P = \left(\frac{\partial F}{\partial V}\right)_T. \tag{A.8}$$

The free energy (F) of solids can be exhibit as the series expansion:

$$F = \sum_{n=1}^{\infty} a_n \epsilon^n \qquad (A.9)$$

where a_n is pressure-dependent coefficient and ϵ^n is the Eulerian strain defined as:

$$\epsilon = \frac{1}{2}\left[1 - \left(\frac{V}{V_0}\right)^{-\frac{2}{3}}\right]. \qquad (A.10)$$

Expansion of series (Equation A.9) yields [128]:

$$P = \frac{3}{2}B_0 \left[\left(\frac{V}{V_0}\right)^{-\frac{7}{3}} - \left(\frac{V}{V_0}\right)^{-\frac{5}{3}}\right]\left[1 + \frac{3}{4}(B'_0 - 4)\left[\left(\frac{V}{V_0}\right)^{-\frac{2}{3}} - 1\right]\right] \qquad (A.11)$$

Bibliography

Bibliography

[1] V. Schmidt, J. V. Wittemann, and U. Gösele, Chemical Reviews **110**, 361 (2010), ISSN 1520-6890, URL http://www.ncbi.nlm.nih.gov/pubmed/20070117.

[2] Z. Wang and J. L. Coffer, The Journal of Physical Chemistry B **108**, 2497 (2004), ISSN 1520-6106, URL http://pubs.acs.org/doi/abs/10.1021/jp036170q.

[3] Z. Wang and J. L. Coffer, Nano Letters **2**, 1303 (2002), ISSN 1530-6984, URL http://pubs.acs.org/doi/abs/10.1021/nl025771a.

[4] X. Zhao, C. M. Wei, L. Yang, and M. Y. Chou, Physical Review Letters **92**, 236805 (2004), ISSN 0031-9007, URL http://link.aps.org/doi/10.1103/PhysRevLett.92.236805.

[5] D. D. D. Ma, C. S. Lee, F. C. K. Au, S. Y. Tong, and S. T. Lee, Science (New York, N.Y.) **299**, 1874 (2003), ISSN 1095-9203, URL http://www.ncbi.nlm.nih.gov/pubmed/12595610.

[6] J. Huo, R. Solanki, J. L. Freeouf, and J. R. Carruthers, Nanotechnology **15**, 1848 (2004), ISSN 0957-4484, URL http://stacks.iop.org/0957-4484/15/i=12/a=027?key=crossref.ad8d4280ae017720f9685f390f6f5763.

[7] X. Han, K. Zheng, Y. Zhang, X. Zhang, Z. Zhang, and Z. Wang, Advanced Materials **19**, 2112 (2007), ISSN 09359648, URL http://doi.wiley.com/10.1002/adma.200602705.

BIBLIOGRAPHY

[8] X. Li, T. Ono, Y. Wang, and M. Esashi, Applied Physics Letters **83**, 3081 (2003), ISSN 00036951, URL http://link.aip.org/link/APPLAB/v83/i15/p3081/s1&Agg=doi.

[9] T. Kizuka, Y. Takatani, K. Asaka, and R. Yoshizaki, Physical Review B **72**, 1 (2005), ISSN 1098-0121, URL http://link.aps.org/doi/10.1103/PhysRevB.72.035333.

[10] M. Tabib-Azar, M. Nassirou, R. Wang, S. Sharma, T. I. Kamins, M. S. Islam, and R. S. Williams, Applied Physics Letters **87**, 113102 (2005), ISSN 00036951, URL http://link.aip.org/link/APPLAB/v87/i11/p113102/s1&Agg=doi.

[11] A. Debernardi, C. Ulrich, and K. Syassen, Physical Review B **59**, 6774 (1999), URL http://link.aps.org/doi/10.1103/PhysRevB.59.6774.

[12] R. Miller, Nanotechnology **11**, 139 (2000), URL http://iopscience.iop.org/0957-4484/11/3/301.

[13] Z. Wang and J. Coffer, J. Phys. Chem. B **108**, 2497 (2004), URL http://pubs.acs.org/doi/abs/10.1021/jp036170q.

[14] R. P. Wang, G. W. Zhou, Y. L. Liu, S. H. Pan, H. Z. Zhang, D. P. Yu, and Z. Zhang, Physical Review B **61**, 16827 (2000), ISSN 0163-1829, URL http://link.aps.org/doi/10.1103/PhysRevB.61.16827.

[15] Y. Sun and Y. Xia, Science (New York, N.Y.) **298**, 2176 (2002), ISSN 1095-9203, URL http://www.ncbi.nlm.nih.gov/pubmed/12481134.

[16] Y. Yin and A. P. Alivisatos, Nature **437**, 664 (2005), ISSN 1476-4687, URL http://www.ncbi.nlm.nih.gov/pubmed/16193041.

[17] V. F. Puntes, D. Zanchet, C. K. Erdonmez, and A. P. Alivisatos, Journal of the American Chemical Society **124**, 12874 (2002), ISSN 0002-7863, URL http://www.ncbi.nlm.nih.gov/pubmed/12392435.

[18] H. Yu and W. Buhro, Advanced Materials **15**, 416 (2003), ISSN 09359648, URL http://doi.wiley.com/10.1002/adma.200390096.

[19] D. V. Talapin, J. H. Nelson, E. V. Shevchenko, S. Aloni, B. Sadtler, and A. P. Alivisatos, Nano letters **7**, 2951 (2007), ISSN 1530-6984, URL http://www.ncbi.nlm.nih.gov/pubmed/17845068.

[20] L. Manna, E. C. Scher, L.-S. Li, and A. P. Alivisatos, Journal of the American Chemical Society **124**, 7136 (2002), ISSN 0002-7863, URL http://www.ncbi.nlm.nih.gov/pubmed/12059239.

[21] S. Lee, N. Wang, and C. Lee, Materials Science and Engineering A **286**, 16 (2000), ISSN 0921-5093, URL http://linkinghub.elsevier.com/retrieve/pii/S0921509300006584.

[22] S. Bhattacharya, D. Banerjee, K. W. Adu, S. Samui, and S. Bhattacharyya, Applied Physics Letters **85**, 2008 (2004), ISSN 00036951, URL http://link.aip.org/link/APPLAB/v85/i11/p2008/s1&Agg=doi.

[23] P. Servati, A. Colli, S. Hofmann, Y. Fu, P. Beecher, Z. Durrani, A. C. Ferrari, a. Flewitt, J. Robertson, and W. Milne, Physica E: Low-dimensional Systems and Nanostructures **38**, 64 (2007), ISSN 13869477, URL http://linkinghub.elsevier.com/retrieve/pii/S138694770600590X.

[24] V. A. Sivakov, R. Scholz, F. Syrowatka, F. Falk, U. Gösele, and S. H. Christiansen, Nanotechnology **20**, 405607 (2009), ISSN 0957-4484, URL http://stacks.iop.org/0957-4484/20/i=40/a=405607?key=crossref.9771a2bffe7af57b29cac02a11afef4c.

[25] T. Stelzner, M. Pietsch, G. Andrä, F. Falk, E. Ose, and S. Christiansen, Nanotechnology **19**, 295203 (2008), ISSN 0957-4484, URL http://stacks.iop.org/0957-4484/19/i=29/a=295203?key=crossref.c51e4487c6647632fc87bd4ac1f623b7.

[26] M. Jeon and K. Kamisako, Materials Letters **63**, 777 (2009), ISSN 0167-577X, URL http://www.sciencedirect.com/science/article/B6TX9-4VB5JXS-1/2/ce4deb4d83a92fb3e08c91ca9787b88c.

[27] L. Hicks and M. Dresselhaus, Physical review B **47**, 16631 (1993), ISSN 1550-235X, URL http://link.aps.org/doi/10.1103/PhysRevB.47.16631.

BIBLIOGRAPHY

[28] Y. Zhang, J. Christofferson, A. Shakouri, D. Li, A. Majumdar, Y. Wu, R. Fan, and P. Yang, Nanotechnology, IEEE Transactions on **5**, 67 (2006), ISSN 1536-125X, URL http://ieeexplore.ieee.org/xpls/abs_all.jsp?arnumber=1576739.

[29] S. Piscanec, M. Cantoro, A. Ferrari, J. Zapien, Y. Lifshitz, S. Lee, S. Hofmann, and J. Robertson, Physical Review B **68**, 2 (2003), ISSN 0163-1829, URL http://link.aps.org/doi/10.1103/PhysRevB.68.241312.

[30] S. Khachadorian, H. Scheel, M. Cantoro, A. Colli, A. C. Ferrari, and C. Thomsen, physica status solidi (b) **246**, 2809 (2009), ISSN 03701972, URL http://doi.wiley.com/10.1002/pssb.200982341.

[31] H. Scheel, S. Khachadorian, M. Cantoro, A. Colli, A. C. Ferrari, and C. Thomsen, Physica Status Solidi (B) **245**, 2090 (2008), ISSN 03701972, URL http://doi.wiley.com/10.1002/pssb.200879554.

[32] H. Scheel, S. Reich, A. C. Ferrari, M. Cantoro, A. Colli, and C. Thomsen, Applied Physics Letters **88**, 233114 (2006), ISSN 00036951, URL http://link.aip.org/link/APPLAB/v88/i23/p233114/s1&Agg=doi.

[33] A. Colli, A. Fasoli, C. Ronning, S. Pisana, S. Piscanec, and A. C. Ferrari, Nano letters **8**, 2188 (2008), ISSN 1530-6984, URL http://www.ncbi.nlm.nih.gov/pubmed/18576693.

[34] S. Hofmann, C. Ducati, R. J. Neill, S. Piscanec, A. C. Ferrari, J. Geng, R. E. Dunin-Borkowski, and J. Robertson, Journal of Applied Physics **94**, 6005 (2003), ISSN 00218979, URL http://link.aip.org/link/JAPIAU/v94/i9/p6005/s1&Agg=doi.

[35] A. Colli, A. Fasoli, P. Beecher, P. Servati, S. Pisana, Y. Fu, A. J. Flewitt, W. I. Milne, J. Robertson, C. Ducati, et al., Journal of Applied Physics **102**, 034302 (2007), ISSN 00218979, URL http://link.aip.org/link/JAPIAU/v102/i3/p034302/s1&Agg=doi.

[36] H. Kohno, T. Iwasaki, Y. Mita, and S. Takeda, Journal of Applied Physics **91**, 3232 (2002), ISSN 00218979, URL http://link.aip.org/link/JAPIAU/v91/i5/p3232/s1&Agg=doi.

[37] J. Qi, Chemical Physics Letters **372**, 763 (2003), ISSN 00092614, URL http://linkinghub.elsevier.com/retrieve/pii/S0009261403005049.

[38] H. Richter, Z. Wang, and L. Ley, Solid State Communications **39**, 625 (1981), ISSN 0038-1098, URL http://linkinghub.elsevier.com/retrieve/pii/0038109881903379.

[39] I. Campbell and P. Fauchet, Solid State Communications **58**, 739 (1986), ISSN 0038-1098, URL http://linkinghub.elsevier.com/retrieve/pii/0038109886905132.

[40] M. Hanfland, U. Schwarz, K. Syassen, and K. Takemura, Physical Review Letters **82**, 1197 (1999), ISSN 0031-9007, URL http://link.aps.org/doi/10.1103/PhysRevLett.82.1197.

[41] *Semiconductors - Basic Data (Berlin, Heidelberg: Springer)* (1996).

[42] B. M. Kayes, M. A. Filler, M. C. Putnam, M. D. Kelzenberg, N. S. Lewis, and H. A. Atwater, Applied Physics Letters **91**, 103110 (2007), ISSN 00036951, URL http://link.aip.org/link/APPLAB/v91/i10/p103110/s1&Agg=doi.

[43] A. I. Hochbaum, R. Fan, R. He, and P. Yang, Nano letters **5**, 457 (2005), ISSN 1530-6984, URL http://www.ncbi.nlm.nih.gov/pubmed/15755094.

[44] S. Hofmann, C. Ducati, R. J. Neill, S. Piscanec, A. C. Ferrari, J. Geng, R. E. Dunin-Borkowski, and J. Robertson, Journal of Applied Physics **94**, 6005 (2003), ISSN 00218979, URL http://link.aip.org/link/JAPIAU/v94/i9/p6005/s1&Agg=doi.

[45] X. Zeng, Journal of Crystal Growth **247**, 13 (2003), ISSN 00220248, URL http://linkinghub.elsevier.com/retrieve/pii/S0022024802019012.

BIBLIOGRAPHY

[46] S.-W. Cheng and H.-F. Cheung, Journal of Applied Physics **94**, 1190 (2003), ISSN 00218979, URL http://link.aip.org/link/JAPIAU/v94/i2/p1190/s1&Agg=doi.

[47] O. Englander, D. Christensen, J. Kim, L. Lin, and S. J. S. Morris, Nano letters **5**, 705 (2005), ISSN 1530-6984, URL http://www.ncbi.nlm.nih.gov/pubmed/15826112.

[48] Z. W. Pan, Z. R. Dai, L. Xu, S. T. Lee, and Z. L. Wang, The Journal of Physical Chemistry B **105**, 2507 (2001), ISSN 1520-6106, URL http://pubs.acs.org/doi/abs/10.1021/jp004253q.

[49] A. Colli, A. Fasoli, P. Beecher, P. Servati, S. Pisana, Y. Fu, A. J. Flewitt, W. I. Milne, J. Robertson, C. Ducati, et al., Journal of Applied Physics **102**, 034302 (2007), ISSN 00218979, URL http://link.aip.org/link/JAPIAU/v102/i3/p034302/s1&Agg=doi.

[50] M. McMahon, R. Nelmes, N. Wright, and D. Allan, Physical Review B **50**, 739 (1994), ISSN 0163-1829, URL http://link.aps.org/doi/10.1103/PhysRevB.50.739.

[51] D. P. Yu, Z. G. Bai, Y. Ding, Q. L. Hang, H. Z. Zhang, J. J. Wang, Y. H. Zou, W. Qian, G. C. Xiong, H. T. Zhou, et al., Applied Physics Letters **72**, 3458 (1998), ISSN 00036951, URL http://link.aip.org/link/APPLAB/v72/i26/p3458/s1&Agg=doi.

[52] Z. Su, J. Sha, G. Pan, J. Liu, D. Yang, C. Dickinson, and W. Zhou, The journal of physical chemistry. B **110**, 1229 (2006), ISSN 1520-6106, URL http://www.ncbi.nlm.nih.gov/pubmed/16471668.

[53] Y. Wang, J. Zhang, J. Wu, J. L. Coffer, Z. Lin, S. V. Sinogeikin, W. Yang, and Y. Zhao, Nano letters **8**, 2891 (2008), ISSN 1530-6984, URL http://www.ncbi.nlm.nih.gov/pubmed/18720974.

[54] M. A. Rafiq, Z. A. K. Durrani, H. Mizuta, A. Colli, P. Servati, A. C. Ferrari, W. I. Milne, and S. Oda, Journal of Applied Physics **103**, 053705

(2008), ISSN 00218979, URL http://link.aip.org/link/JAPIAU/v103/i5/p053705/s1&Agg=doi.

[55] R.-Q. Zhang, Y. Lifshitz, and S.-T. Lee, Advanced Materials **15**, 635 (2003), ISSN 09359648, URL http://doi.wiley.com/10.1002/adma.200301641.

[56] R. Tubino, The Journal of Chemical Physics **56**, 1022 (1972), ISSN 00219606, URL http://link.aip.org/link/?JCP/56/1022/1&Agg=doi.

[57] G. Nilsson and G. Nelin, Physical Review B **6**, 3777 (1972), ISSN 0556-2805, URL http://link.aps.org/doi/10.1103/PhysRevB.6.3777.

[58] A. Heidelberg, L. T. Ngo, B. Wu, M. A. Phillips, S. Sharma, T. I. Kamins, J. E. Sader, and J. J. Boland, Nano letters **6**, 1101 (2006), ISSN 1530-6984, URL http://www.ncbi.nlm.nih.gov/pubmed/16771561.

[59] M. Gordon, T. Baron, F. Dhalluin, and P. Gentile, Nano letters pp. 2–6 (2009), URL http://pubs.acs.org/doi/abs/10.1021/nl802556d.

[60] E. Anastassakis, Solid State Communications **8**, 133 (1970), ISSN 00381098, URL http://linkinghub.elsevier.com/retrieve/pii/0038109870905880.

[61] F. Cerdeira, C. Buchenauer, F. Pollak, and M. Cardona, Physical Review B **5**, 580 (1972), ISSN 0556-2805, URL http://link.aps.org/doi/10.1103/PhysRevB.5.580.

[62] J. Sandler, M. Shaffer, A. Windle, M. Halsall, M. Montes-Morán, C. Cooper, and R. Young, Physical Review B **67**, 1 (2003), ISSN 0163-1829, URL http://link.aps.org/doi/10.1103/PhysRevB.67.035417.

[63] G. Lucazeau, Journal of Raman Spectroscopy **34**, 478 (2003), ISSN 0377-0486, URL http://doi.wiley.com/10.1002/jrs.1027.

[64] W. Sherman, Journal of Physics C: Solid State Physics **13**, 4601 (1980), URL http://iopscience.iop.org/0022-3719/13/25/005.

BIBLIOGRAPHY

[65] W. Sherman and G. Wilkinson, *Advances in Infrared and Raman Spectroscopy* (Heydon: London, 1983).

[66] W. C. O'mara, R. B. Herring, and L. P. Hunt, eds., *Handbook of Semiconductor Silicon Technology* (Noyes publications, Mill Road, Park Ridge, New Jersey, 1990).

[67] S. Ganesan, A. Maradudin, and J. Oitmaa, Annals of Physics **56**, 556 (1970), ISSN 00034916, URL http://linkinghub.elsevier.com/retrieve/pii/0003491670900291.

[68] G. Huber, K. Syassen, and W. Holzapfel, Physical Review B **15**, 5123 (1977), ISSN 0556-2805, URL http://link.aps.org/doi/10.1103/PhysRevB.15.5123.

[69] A. Jayaraman, Reviews of Modern Physics (1983), URL http://link.aps.org/doi/10.1103/RevModPhys.55.65.

[70] R. A. Forman, G. J. Piermarini, J. D. Barnett, and S. Block, Science (New York, N.Y.) **176**, 284 (1972), ISSN 0036-8075, URL http://www.ncbi.nlm.nih.gov/pubmed/17791916.

[71] D. Li, Y. Wu, P. Kim, L. Shi, P. Yang, and A. Majumdar, Applied Physics Letters **83**, 2934 (2003), ISSN 00036951, URL http://link.aip.org/link/APPLAB/v83/i14/p2934/s1&Agg=doi.

[72] N. Mingo, Physical Review B **68**, 1 (2003), ISSN 1098-0121, URL http://link.aps.org/doi/10.1103/PhysRevB.68.113308.

[73] I. Ponomareva, D. Srivastava, and M. Menon, Nano letters **7**, 1155 (2007), ISSN 1530-6984, URL http://www.ncbi.nlm.nih.gov/pubmed/17394370.

[74] A. I. Boukai, Y. Bunimovich, J. Tahir-Kheli, J.-K. Yu, W. A. Goddard, and J. R. Heath, Nature **451**, 168 (2008), ISSN 1476-4687, URL http://www.ncbi.nlm.nih.gov/pubmed/18185583.

BIBLIOGRAPHY

[75] H. Scheel, S. Reich, A. C. Ferrari, M. Cantoro, A. Colli, and C. Thomsen, Applied Physics Letters **88**, 233114 (2006), ISSN 00036951, URL http://link.aip.org/link/APPLAB/v88/i23/p233114/s1&Agg=doi.

[76] K. Adu, H. Gutiérrez, U. Kim, and P. Eklund, Physical Review B **73**, 1 (2006), ISSN 1098-0121, URL http://link.aps.org/doi/10.1103/PhysRevB.73.155333.

[77] R. Gupta, Q. Xiong, C. K. Adu, U. J. Kim, and P. C. Eklund, Nano Letters **3**, 627 (2003), ISSN 1530-6984, URL http://pubs.acs.org/doi/abs/10.1021/nl0341133.

[78] M. J. Assael, E. Charitidou, and C. A. Nieto De Castro, International Journal of Thermophysics **9**, 813 (1988), ISSN 0195-928X, URL http://www.springerlink.com/index/10.1007/BF00503247.

[79] D. E. Gray, *American Institute of Physics handbook. Section editors, Albert A. Bennett [and others] Coordinating editor: Dwight E. Gray* (McGraw-Hill, New York, 1957), URL http://nla.gov.au/nla.cat-vn712224.

[80] D. Y. Hugh and F. R. A., *University Physics* (Addison Wesley Longman, Inc., 1996).

[81] B. Weinstein and G. Piermarini, Physical Review B **12**, 1172 (1975), ISSN 0556-2805, URL http://link.aps.org/doi/10.1103/PhysRevB.12.1172.

[82] R. Trommer, H. Müller, M. Cardona, and P. Vogl, Physical Review B **21**, 4869 (1980), ISSN 0163-1829, URL http://link.aps.org/doi/10.1103/PhysRevB.21.4869.

[83] J. Hu and I. Spain, Solid State Communications **51**, 263 (1984), ISSN 0038-1098, URL http://linkinghub.elsevier.com/retrieve/pii/0038109884906835.

[84] J. Hu, L. Merkle, C. Menoni, and I. Spain, Physical Review B **34**, 4679 (1986), ISSN 1550-235X, URL http://link.aps.org/doi/10.1103/PhysRevB.34.4679.

BIBLIOGRAPHY

[85] T. Mernagh and L.-G. Liu, Journal of Physics and Chemistry of Solids **52**, 507 (1991), ISSN 00223697, URL http://linkinghub.elsevier.com/retrieve/pii/002236979190183Z.

[86] M. Cohen, Physica Scripta **T1**, 5 (1982), URL http://iopscience.iop.org/1402-4896/1982/T1/001.

[87] M. Cohen, Physical Review B **32**, 7988 (1985), ISSN 0163-1829, URL http://link.aps.org/doi/10.1103/PhysRevB.32.7988.

[88] P. Lam and M. Cohen, Physical Review B (1987), URL http://link.aps.org/doi/10.1103/PhysRevB.35.9190.

[89] A. Debernardi and S. Baroni, Physical review letters **75**, 1819 (1995), ISSN 0031-9007, URL http://link.aps.org/doi/10.1103/PhysRevLett.75.1819.

[90] J. Menendez and M. Cardona, Physical Review B **29**, 2051 (1984), ISSN 0163-1829, URL http://link.aps.org/doi/10.1103/PhysRevB.29.2051.

[91] D. W. Posener, Aust. J. Phys. **12**, 184 (1959).

[92] M. Balkanski, R. Wallis, and E. Haro, Physical Review B **28**, 1928 (1983), ISSN 0163-1829, URL http://link.aps.org/doi/10.1103/PhysRevB.28.1928.

[93] B. Li, D. Yu, and S.-L. Zhang, Physical Review B **59**, 1645 (1999), ISSN 0163-1829, URL http://link.aps.org/doi/10.1103/PhysRevB.59.1645.

[94] F. Galeener and G. Lucovsky, Physical Review Letters **37**, 1474 (1976), ISSN 0031-9007, URL http://link.aps.org/doi/10.1103/PhysRevLett.37.1474.

[95] S. T. Lee, N. Wang, and C. S. Lee, Material Science and Engineering A **286**, 16 (2003), ISSN 1422-6375.

[96] P. Bruesch, ed., *Phonons: Theory and Experiments II* (Springer, Berlin, Berlin, 1986).

BIBLIOGRAPHY

[97] D. Yu, Solid State Communications **105**, 403 (1998), ISSN 00381098, URL http://linkinghub.elsevier.com/retrieve/pii/S0038109897101430.

[98] N. Fukata, T. Oshima, K. Murakami, T. Kizuka, T. Tsurui, and S. Ito, Applied Physics Letters **86**, 213112 (2005), ISSN 00036951, URL http://link.aip.org/link/APPLAB/v86/i21/p213112/s1&Agg=doi.

[99] M. Lu, Chemical Physics Letters **374**, 542 (2003), ISSN 00092614, URL http://linkinghub.elsevier.com/retrieve/pii/S0009261403007474.

[100] A. Lugstein, M. Steinmair, Y. J. Hyun, G. Hauer, P. Pongratz, and E. Bertagnolli, Nano letters **8**, 2310 (2008), ISSN 1530-6984, URL http://www.ncbi.nlm.nih.gov/pubmed/18624392.

[101] B. P., ed., *Phonons: Theory and Experiments I* (Springer, Berlin, Berlin, 1982).

[102] L. gun Liu, Mechanics of Materials **14**, 283 (1993), ISSN 0167-6636, URL http://www.sciencedirect.com/science/article/B6TX6-4811T84-C/2/cbc735fd9defda4031bc1f5d27d0cf0a.

[103] C. R. Jaeger, ed., *Introduction to Microelectronic Fabrication* (Prentice Hall, Upper Saddle River, New Jersey, 1993), second edition ed.

[104] M. Winkelnkemper, Diplomarbeit, *Elektronische Eigenschaften niederdimensionaler Halbleiterstrukturen mit Wurtzitstruktur* (2004).

[105] J. Niu, J. Sha, and D. Yang, Scripta Materialia **55**, 183 (2006), ISSN 13596462, URL http://linkinghub.elsevier.com/retrieve/pii/S1359646206002727.

[106] S. Piscanec, A. C. Ferrari, M. Cantoro, S. Hofmann, J. A. Zapien, Y. Lifshitz, S. T. Lee, and J. Robertson, Materials Science and Engineering: C **23**, 931 (2003), ISSN 09284931, URL http://linkinghub.elsevier.com/retrieve/pii/S092849310300170X.

[107] T. Hart, R. Aggarwal, and B. Lax, Physical Review B **1**, 638 (1970), ISSN 1550-235X, URL http://link.aps.org/doi/10.1103/PhysRevB.1.638.

BIBLIOGRAPHY

[108] R. Tsu and J. Hernandez, Applied Physics Letters **41**, 1016 (2009), ISSN 0003-6951, URL http://ieeexplore.ieee.org/xpls/abs_all.jsp?arnumber=4851068.

[109] R. Tsu, Applied Physics Letters **41**, 1016 (1982), ISSN 00036951, URL http://link.aip.org/link/?APL/41/1016/1&Agg=doi.

[110] P. Klemens, Physical Review **148**, 845 (1966), ISSN 0031-899X, URL http://link.aps.org/doi/10.1103/PhysRev.148.845.

[111] I. Ipatova, A. Maradudin, and R. Wallis, Physical Review **155**, 882 (1967), ISSN 0031-899X, URL http://link.aps.org/doi/10.1103/PhysRev.155.882.

[112] J. B. Cui, K. Amtmann, J. Ristein, and L. Ley, Journal of Applied Physics **83**, 7929 (1998), ISSN 00218979, URL http://link.aip.org/link/JAPIAU/v83/i12/p7929/s1&Agg=doi.

[113] L. Vina, S. Logothetidis, and M. Cardona, Physical Review B **30**, 1979 (1984), ISSN 0163-1829, URL http://link.aps.org/doi/10.1103/PhysRevB.30.1979.

[114] P. Mishra and K. Jain, Physical Review B **62**, 14790 (2000), ISSN 0163-1829, URL http://link.aps.org/doi/10.1103/PhysRevB.62.14790.

[115] P. Temple and C. Hathaway, Physical Review B **7**, 3685 (1973), ISSN 0556-2805, URL http://link.aps.org/doi/10.1103/PhysRevB.7.3685.

[116] K. Uchinokura, Journal of Physics and Chemistry of Solids **35**, 171 (1974), ISSN 00223697, URL http://linkinghub.elsevier.com/retrieve/pii/0022369774900316.

[117] G. Nelin and G. Nilsson, Physical Review B **5**, 3151 (1972), ISSN 0556-2805, URL http://link.aps.org/doi/10.1103/PhysRevB.5.3151.

[118] J. B. Cui, K. Amtmann, J. Ristein, and L. Ley, Journal of Applied Physics **83**, 7929 (1998), ISSN 00218979, URL http://link.aip.org/link/JAPIAU/v83/i12/p7929/s1&Agg=doi.

BIBLIOGRAPHY

[119] V. V. Brazhkin, S. G. Lyapin, I. A. Trojan, R. N. Voloshin, A. G. Lyapin, and N. N. Mel'nik, Journal of Experimental and Theoretical Physics Letters **72**, 195 (2000), ISSN 0021-3640, URL http://www.springerlink.com/index/10.1134/1.1320111.

[120] P. Mishra and K. Jain, Physical Review B **64**, 1 (2001), ISSN 0163-1829, URL http://link.aps.org/doi/10.1103/PhysRevB.64.073304.

[121] W. Dash and R. Newman, Physical Review **99**, 1151 (1955), ISSN 0031-899X, URL http://link.aps.org/doi/10.1103/PhysRev.99.1151.

[122] L. Woods and G. Mahan, Physical Review B **57**, 7679 (1998), ISSN 0163-1829, URL http://link.aps.org/doi/10.1103/PhysRevB.57.7679.

[123] H. Lange, M. Mohr, M. Artemyev, U. Woggon, and C. Thomsen, Nano letters **8**, 4614 (2008), ISSN 1530-6992, URL http://www.ncbi.nlm.nih.gov/pubmed/18983202.

[124] Y. Li, Dissertation, *First-principles Calculations on the Electronic, Vibrational, and Optical Properties of Semiconductor Nanowires* (2006).

[125] X. B. Zeng, X. B. Liao, S. T. Dai, B. Wang, Y. Y. Xu, X. B. Xiang, Z. H. Hu, H. W. Diao, and G. L. Kong, Journal of Metastable and Nanocrystalline Materials **23**, 137 (2005), ISSN 1422-6375, URL http://www.scientific.net/JMNM.23.137.

[126] F. Murnaghan, Proceedings of the national academy of sciences of the United States of America **30**, 244 (1944), URL http://www.ncbi.nlm.nih.gov/pmc/articles/PMC1078704/.

[127] F. Birch, Physical Review **71**, 809 (1947), ISSN 0031-899X, URL http://link.aps.org/doi/10.1103/PhysRev.71.809.

[128] P. Vinet, J. Ferrante, J. Smith, and J. Rose, Journal of Physics C: Solid State Physics **19**, L467 (1986), URL http://iopscience.iop.org/0022-3719/19/20/001.

Acknowledgments

This work has been conducted with the help of others. Especially, I want to thank

C. Thomsen for his valued support and continuous guidance.

K. Papagelis for constructive cooperation and advices.

H. Scheel for the valuable discussions and many advices.

A. Schliwa for the fruitful discussions and the work on the linear elasticity continuum model.

S. Selve for his assistance with TEM measurements.

M. Abramian and N. Peica for support and proofreading.

H. Lange, B. Jeschke and T. Switaiski for proofreading.

Die VDM Verlagsservicegesellschaft sucht für wissenschaftliche Verlage abgeschlossene und herausragende

Dissertationen, Habilitationen, Diplomarbeiten, Master Theses, Magisterarbeiten usw.

für die kostenlose Publikation als Fachbuch.

Sie verfügen über eine Arbeit, die hohen inhaltlichen und formalen Ansprüchen genügt, und haben Interesse an einer honorarvergüteten Publikation?

Dann senden Sie bitte erste Informationen über sich und Ihre Arbeit per Email an *info@vdm-vsg.de*.

Sie erhalten kurzfristig unser Feedback!

VDM Verlagsservicegesellschaft mbH
Dudweiler Landstr. 99   Telefon +49 681 3720 174
D - 66123 Saarbrücken   Fax   +49 681 3720 1749

www.vdm-vsg.de

Die VDM Verlagsservicegesellschaft mbH vertritt

Printed by Books on Demand GmbH, Norderstedt / Germany